Mastering the Rules of Competitive Strategy

OTHER AUERBACH PUBLICATIONS

Ad Hoc Mobile Wireless Networks: Principles, Protocols and Applications
Subir Kumar Sarkar, T.G. Basavaraju and C. Puttamadappa
ISBN 1-4200-6221-2

Computer Fraud: An In-depth Framework for Detecting and Defending against Insider IT Attacks
Kenneth C. Brancik
ISBN 1-4200-4659-4

Design Science Research Methods and Patterns: Innovating Information and Communication Technology
Vijay K. Vaishnavi and William Kuechler Jr.
ISBN 1-4200-5932-7

Determining Project Requirements
Hans Jonasson
ISBN 1-4200-4502-4

Digital Privacy: Theory, Technologies, and Practices
Alessandro Acquisti, Stefanos Gritzalis, Costas Lambrinoudakis and Sabrina De Capitani di Vimercati
ISBN 1-4200-521-79

Effective Communications for Project Management
Ralph L. Kliem
ISBN 1-4200-6246-8

Effective Transition from Design to Production
David F. Ciambrone
ISBN 1-4200-4686-1

Elements of Compiler Design
Alexander Meduna
ISBN 1-4200-6323-5

How to Achieve 27001 Certification: An Example of Applied Compliance Management
Sigurjon Thor Arnason and Keith D. Willett
ISBN 0-8493-3648-1

Inter- and Intra-Vehicle Communications
Gilbert Held
ISBN 1-4200-5221-7

Manage Software Testing
Peter Farrell-Vinay
ISBN 0-8493-9383-3

Managing Global Development Risk
James M. Hussey and Steven E. Hall
ISBN 1-4200-5520-8

Mobile WiMAX: Toward Broadband Wireless Metropolitan Area Networks
Yan Zhang and Hsiao-Hwa Chen
ISBN 0-8493-2624-9

Operational Excellence: Using Lean Six Sigma to Translate Customer Value through Global Supply Chain
James William Martin
ISBN 1-4200-6250-6

Physical Principles of Wireless Communications
Victor L. Granatstein
ISBN 0-8493-3259-1

Practical Guide to Project Planning
Ricardo Viana Vargas
ISBN 1-4200-4504-0

Principles of Mobile Computing and Communications
Mazliza Othman
ISBN 1-4200-6158-5

Programming Languages for Business Problem Solving Price
Shouhong Wang and Hai Wang
ISBN 1-4200-6264-6

Retail Supply Chain Management
James B. Ayers and Mary Ann Odegaard
ISBN 0-8493-9052-4

Security in Wireless Mesh Networks
Yan Zhang, Jun Zheng and Honglin Hu
ISBN 0-8493-8250-5

Service-Oriented Architecture: SOA Strategy, Methodology, and Technology
James P. Lawler and H. Howell-Barber
ISBN 1-4200-4500-8

The Strategic Project Leader: Mastering Service-Based Project Leadership
Jack Ferraro
ISBN 0-8493-8794-9

Simplified TRIZ: New Problem Solving Applications for Engineers and Manufacturing Professionals, Second Edition
Kalevi Rantanen and Ellen Domb
ISBN 1-4200-6273-5

Value-Added Services for Next Generation Networks
Thierry Van de Velde
ISBN 0-8493-7318-2

WiMAX: A Wireless Technology Revolution
G.S.V. Radha Krishna Rao and G. Radhamani
ISBN 0-8493-7059-0

AUERBACH PUBLICATIONS
www.auerbach-publications.com
To Order Call: 1-800-272-7737 • Fax: 1-800-374-3401
E-mail: orders@crcpress.com

Mastering the Rules of Competitive Strategy

A Resource Guide for Managers

NORTON PALEY

Auerbach Publications
Taylor & Francis Group
New York London

CRC Press is an imprint of the
Taylor & Francis Group, an **informa** business

Auerbach Publications
Taylor & Francis Group
6000 Broken Sound Parkway NW, Suite 300
Boca Raton, FL 33487-2742

© 2008 by Taylor & Francis Group, LLC
Auerbach is an imprint of Taylor & Francis Group, an Informa business

No claim to original U.S. Government works
Printed in the United States of America on acid-free paper
10 9 8 7 6 5 4 3 2 1

International Standard Book Number-13: 978-1-4200-6809-2 (Hardcover)

This book contains information obtained from authentic and highly regarded sources. Reprinted material is quoted with permission, and sources are indicated. A wide variety of references are listed. Reasonable efforts have been made to publish reliable data and information, but the author and the publisher cannot assume responsibility for the validity of all materials or for the consequences of their use.

Except as permitted under U.S. Copyright Law, no part of this book may be reprinted, reproduced, transmitted, or utilized in any form by any electronic, mechanical, or other means, now known or hereafter invented, including photocopying, microfilming, and recording, or in any information storage or retrieval system, without written permission from the publishers.

For permission to photocopy or use material electronically from this work, please access www.copyright.com (http://www.copyright.com/) or contact the Copyright Clearance Center, Inc. (CCC) 222 Rosewood Drive, Danvers, MA 01923, 978-750-8400. CCC is a not-for-profit organization that provides licenses and registration for a variety of users. For organizations that have been granted a photocopy license by the CCC, a separate system of payment has been arranged.

Trademark Notice: Product or corporate names may be trademarks or registered trademarks, and are used only for identification and explanation without intent to infringe.

Library of Congress Cataloging-in-Publication Data

Paley, Norton.
   Mastering the rules of competitive strategy : a resource guide for managers / Norton Paley.
      p. cm.
   Includes bibliographical references and index.
   ISBN 978-1-4200-6809-2 (hbk. : alk. paper) 1. Marketing--Management. 2. Strategic planning--Management. I. Title.

HF5415.13.P322 2008
658.8--dc22   2007034142

Visit the Taylor & Francis Web site at
http://www.taylorandfrancis.com

and the Auerbach Web site at
http://www.auerbach-publications.com

Contents

Dedication xi
Introduction xiii
    Defining Strategy for Business Applications xvi
        Roots of Strategy xvii
    Organization of the Book xviii
The Author xxi

1   **Shift to the Offense: Turn a Risky Competitive Situation into a Fresh Market Opportunity** 1
    Shift to the Offensive with Boldness 3
    Make Reliable Estimates and Calculations 4
        Internal and External Relationships 4
        Seasonal Forces 6
        Market Selection 6
        Policy 7
        Prioritize Your Actions 8
    Hold Reserves to Exploit Opportunities 10
    Assess Levels of Creativity and Innovation 12
    Evaluate the Ability of Personnel to Deal with Friction 14
    Limit the Negative Effects of Friction 15
    Strategy Diagnostic Tool 16
        Strategy Rule 1: Shift to the Offensive 16

2   **Maneuver by Indirect Strategy: Apply Strength against Weakness** 19
    The Power of Indirect Strategy 20
    Developing an Indirect Strategy 22
        Define the Strategic Goals 22
            What Are Our Organization's Distinctive Areas of Expertise? 22
            What Should Be Our Organization's Business during the Next Three to Five Years? 23
            What Segments or Categories of Customers Will We Serve? 23

What Additional Functions Are We Likely to Fulfill for
Customers as We See the Market Evolve? 24
What New Technologies Will Our Firm Require to Satisfy
Future Customer Needs? 24
What Changes Taking Place in Markets, Consumer Behavior,
Competition, Environment, Culture, and the Economy Will
Impact Our Company? 24
Determine the Resources Needed to Achieve the Strategic Goals 26
Gather Competitive Intelligence 28
Establish Security 30
Implement the Strategy 32
Develop a Poststrategy 33
Prolonging the Life of Existing Products and Creating New
Product Opportunities 34
Enhancing and Prolonging Customer Relationships 35
Summary 36
Strategy Diagnostic Tool 36
Strategy Rule 2: Maneuver by Indirect Strategy 36

3 Act with Speed: The Essential Component to Secure a Competitive Lead 39
Strategic Value of Speed 40
Barriers to Speed 43
Lack of Reliable Market Intelligence 44
Mediocre Leadership Stifles Timely and Significant Progress 45
A Manager's Low Self-Esteem and Indecisiveness as Deep-Rooted
Personality Traits 46
Lack of Courage to Go on the Offensive, Triggered by a Manager's
Innate Fear of Failure 46
No Trust by the Manager in Employees' Discipline, Capabilities,
or Skills 47
No Trust by Employees in Their Manager's Ability to Make
Correct Decisions 47
Inadequate Support from Senior Management 48
Disagreement and Open Confrontations among Line Managers
about Objectives, Priorities, and Strategies 48
A Highly Conservative and Plodding Corporate Culture Places a
Drag on Speed 49
Lack of Urgency in Developing New Products to Deal with Short
Product Life Cycles 49
Organizational Layers, Long Chains of Command, and
Cumbersome Committees Prolong Deliberation and Foster
Procrastination 49

Aggressive Competitors Can Strike Fear among Employees,
Damage Morale, and Result in Lost Momentum 50
Complacency or Arrogance as a Prevailing Cultural Mindset 52
Speed: A Core Rule of Strategy...52
Summary..53
Strategy Diagnostic Tool...53
Strategy Rule 3: Act with Speed 53

4 Grow by Concentration: Deploy, Target, Segment.............................57
Select Your Segment and Shape Your Competition 64
Guidelines to Utilizing Grassroots Ethnography.......................................67
Step 1: Map a Segment..67
Step 2: Create a Special Language 68
Step 3: Observe Body Language 68
Step 4: Describe the Ritual..68
Eight Categories of Markets..69
Natural Markets..69
Leading-Edge Markets 72
Key Markets..73
Linked Markets...73
Central Markets..75
Challenging Markets...76
Difficult Markets...76
Encircled Markets 77
Summary..82
Strategy Diagnostic Tool...83
Strategy Rule 4: Grow by Concentration...83

5 Prioritize Competitive Intelligence: The Underpinnings of
Business Strategy...87
Competitive Intelligence: The Underpinnings of Business Strategy 89
The Call to Action.. 90
Tools and Techniques of Competitive Intelligence....................................92
Sales Force...92
Customer Surveys 92
Published Data..93
Government Agencies 93
Industry Studies..94
On-Site Observations 94
Competitive Benchmarking 94
Internal Competitive Intelligence 95
Reverse Engineering..96
Market Signals 97

Contents

- Employing Agents 98
  - Native Agents 98
  - Inside Agents 99
  - Double Agents 100
  - Expendable Agents 100
  - Living Agents 100
- Competitive Intelligence Applications 101
  - Market Selection 101
  - Maneuvering 102
  - Positioning 103
- Summary 104
- Strategy Diagnostic Tool 105
  - Strategy Rule 5: Prioritize Competitive Intelligence 105

6 Align Competitive Strategy with Your Corporate Culture: The Lifeline to Your Organization's Future 107

- Characteristics of High-Performing Business Cultures 110
- Decoding Competitors 117
  - Beliefs 119
  - Symbols and Rituals 122
- Energize Your Company's Culture 124
- Summary 126
- Strategy Diagnostic Tool 127
  - Strategy Rule 6: Align Competitive Strategy with Your Corporate Culture 127

7 Develop Leadership Skills: The Moral Fiber Underlying Business Strategy 129

- Self-Confidence 134
- Qualities of Successful Leaders 135
  - Levels of Leadership 135
- Leadership in the Competitive Arena 140
  - Leadership and Implementing Strategy 142
- Power Up Your Business Strategy 143
- Strategy Diagnostic Tool 145
  - Strategy Rule 7: Develop Leadership Skills 145

8 Create a Morale Advantage: Engage Heart, Mind, and Spirit When All Else Fails 149

- Motivational Behavior 152
  - Herzberg's Motivation-Hygiene Theory 152
  - McGregor's X and Y Theories 152

Maslow's Hierarchy of Needs 152
Ouchi's Theory Z 154
Morale Meets Technology 155
Morale Faces the Human Heart 157
Morale Connects with the Power of Unity 158
Activating High Morale and Maintaining Momentum 159
Barriers to Building Morale 161
Action Steps to Build Morale 163
Morale Interfaces with Innovation 164
Morale Links with Trust 166
Summary 166
Strategy Diagnostic Tool 166
Strategy Rule 8: Create a Morale Advantage 166

9 Strengthen Your Decision-Making Capabilities: Fortify Intuition, Enhance Business Experience, Expand Knowledge 171
Valuing Business History 176
Managing Knowledge 178
Activating Intuition 179
The Human Element 184
Strategy Diagnostic Tool 186
Strategy Rule 9: Strengthen Your Decision-Making Capabilities 186

Appendix The Strategic Business Plan: Forms and Guidelines 189
Overview of the Strategic Business Plan Strategic Section 190
Section 1: Strategic Direction 190
Planning Guidelines 191
Section 2: Objectives and Goals 192
Planning Guidelines 192
Quantitative Objectives 193
Nonquantitative Objectives 193
Section 3: Growth Strategies 193
Planning Guidelines 193
Section 4: Business Portfolio Plan 194
Planning Guidelines 194
Overview of the Strategic Business Plan Tactical Section 196
Section 5: Situation Analysis 197
Planning Guidelines 197
Level A: Marketing Mix—Product 197
Level A: Marketing Mix—Pricing 199
Level A: Marketing Mix—Supply Chain and Methods 199

  Level A: Marketing Mix—Advertising, Sales Promotion,
   Internet, and Publicity 200
  Level B: Competitive Analysis 201
  Level C: Market Background 204
Section 6: Marketing Opportunities 206
  Planning Guidelines 206
Section 7: Tactical Objectives 208
  Planning Guidelines 208
   Assumptions 208
   Primary Objectives 209
   Targets of Opportunity Objectives 210
Section 8: Strategies and Tactics 214
  Planning Guidelines 214
  Summary Strategy 215
Section 9: Financial Controls and Budgets 215
  Planning Guidelines 215

Index **217**

Dedication

To Annette: my wife, best friend, and enduring supporter

Introduction

Why is it that some companies overcome huge obstacles and achieve spectacular successes and others are giant flops? Is it mere luck that some senior executives and midlevel managers continue to outperform the market, while others flounder and toil in leftover, barely profitable segments? Do the winning and losing managers not face similar obstacles? Externally, they confront domestic and global competitors, feel the inflationary threats of rising costs and diminishing profits, and struggle to introduce innovative products and services before rivals beat them to market. Internally, they maneuver for executives' attention, compete in an ongoing battle for funds against other product lines or business units, function within mandatory budgetary constraints, and react to the unrelenting demands for increased sales and profits.

The stakes are high and real. Money is spent. Personnel are committed in a contest that means the triumph or downfall of a company. Yet, given all or a few of these barriers, some companies and their executives continue to thrive while others wash out or become also-rans. In the end, however, it is the dynamics of the competitive marketplace, the quality of the organization's business plan, and the ability to implement strategies successfully that validate a company's existence.

The following three "rise then fall, then rise again" examples illustrate these points.

Case Example 1

Intel Corp. created a thriving corporate culture built on a belief that "only the paranoid survive." In a demanding, tightly controlled corporate structure, engineers ruled the company as they developed ever faster microprocessors for PCs and servers that helped Intel bury the competition. Then, the company's phenomenal double-digit growth slowed to a paltry single-digit level. The primary reason was that PC growth weakened as cell phones and a huge array of other handheld devices competed for consumers' attention.

About the same time, the brilliant and legendary CEO Andrew Grove stepped down and the newly minted CEO Paul Otellini took the helm. The transition set in motion a process of phasing out the methods that had worked so well in past decades. Otellini began remaking Intel. He moved beyond microprocessors to create platforms of microprocessors, combining silicon and software that would lead to new devices and technologies and to new markets. He created cross-functional teams of marketers, software developers, and even cultural anthropologists that shared equal authority with engineers.

Next, a new "Leap Ahead" logo replaced the more than 30-year-old "Intel Inside" symbol. With the changeover a resounding message flashed into the marketplace: Watch out. Intel was not only moving, but it was also leaping forward with new technologies, products, and services.

Otellini also expanded from the previous strict and centralized organizational design to a decentralized approachable culture where openness and creativity with a market-driven orientation would rule. Open communication was encouraged with the pervasive use of e-mail as the primary channel to reach employees and prioritize the need for unity.

Case Example 2

Dr. Reddy's Laboratories and **Ranbaxy Laboratories,** leading drug makers in India, have enjoyed phenomenal success by manufacturing and selling generic drugs in North America and Europe. In the process they turned into "Big Pharma's" fiercest competitors by providing consumers with low-cost versions of brand-name drugs such as Prozac and Augmentin. The major pharmaceutical companies that once held the patents to those formulas soon felt the effects of depleting profits.

However, the good times came to a screeching halt for Dr. Reddy's and Ranbaxy as competitors pounced from several directions. First, a wave of small Indian companies entered the fray, selling their generics in the United States and Europe. Then, several unforeseen actions occurred. Major Western drug makers began fighting back. **Pfizer** won a testy and high-profile battle with Ranbaxy over generic rights to its cholesterol drug Lipitor. Other large pharmaceutical companies preempted their Indian rivals by gearing up to sell generic versions of their own blockbuster drugs once patents expired.

How, then, did Dr Reddy's and Ranbaxy cope with the market turmoil? Strategically, each company recognized the urgent need to defend its core business by looking for innovative approaches to sustain a market presence. One workable strategy involved going through the back doors to reach the U.S. and European markets. Both companies cut cooperative marketing deals with midsize, regional pharmaceutical companies. Other actions entailed acquiring smaller generic labs in their respective geographic regions. Although it has been a slow process, the uphill climbs for Dr. Reddy's and Ranbaxy have resulted in growth equal to previous rates of growth.

Case Example 3
Electronic Arts (EA), a video game developer, flourished with a business model that made it a moneymaking leader in its industry. The company created games based on popular sports or founded on movie characters such as Harry Potter, James Bond, and the Lord of the Rings.

As a follow-up, new versions of the games were developed with great frequency, as long as the characters remained trendy. Additionally, to assure successful product launches, top-of-mind identity, profitability, and market leadership, EA spent heavily on marketing. Then the carefully crafted strategy faltered.

The downturn was due primarily to movie studios and sports leagues driving up the costs of licenses. To further aggravate the situation, video game sales in the industry remained sluggish. These burning issues became a wake-up call that forced EA's president, Paul Lee, to change strategies and move quickly before hungry competitors gained dominant positions.

What did a change in strategy mean for EA? First, there was the internal issue of transforming a risk-averse corporate culture known for its operational efficiency into a breeding ground for innovative games. To that end, the company aggressively hired new talent, including movie directors with strong track records for innovation and creativity. Second, EA built a modern-day studio that focused entirely on developing original titles. Here, too, the aim was to drive imaginative products. Electronic Arts also moved rapidly to acquire independent studios. Third, EA worked aggressively and sensitively on ways to boost the staff's morale and competitive spirit. Organizationally, that meant breaking the development staff into small, cohesive, six- to eight-person "cells" and assigning each cell a specific task for the game. This move was in contrast with the former method of rotating individuals, as needed, to projects to meet rigid deadlines.

To further harness team spirit and build morale, each week the teams with the most creative breakthroughs had their work featured on the TV screens that were placed throughout the studio. With quality and customer satisfaction at the forefront of activity, the previous policy of meeting deadlines at all costs has given way to one of "have the best games and release them when they are ready."

What, then, are the commonalities or focal points that turn companies from the brink of failure into a hotbed of winning performance? The following list indicates capabilities that steadily contribute to successful achievements; use it to benchmark areas that would contribute to your company's competitive improvement:

- Shape fresh strategies and tactics so that they are not worn-out repeats of yesterday's action. Your strategies should achieve the dual objectives of securing a market position while defending against aggressive competitors.
- Feed off the hard-earned lessons of past market campaigns. Find out what worked, what did not work, and what could be salvaged for next-generation strategies.
- Align business plans and strategies with your organization's culture. You thereby differentiate your firm and give it a special uniqueness.
- Maintain ongoing competitive intelligence. If you know your competitor's moves, you can determine with some accuracy which strategies will likely succeed and reject those with little chance of success.
- Lift employees' morale and encourage their trust. The intrepid leader is still the decisive component in any endeavor.
- Instill discipline and training to prevent employees from caving in to tough competitive situations. The intent is to avoid excessive caution that can result in loss of equilibrium, initiative, and a winning spirit.
- Foster team solidarity that inspires innovation. For anything significant to happen, the mind and quality of the individual, as well as the unity of team effort, produce a meaningful effect.

Defining Strategy for Business Applications

At this point, it is useful to gain a common understanding of business strategy—a term that is freely tossed around and applied to numerous functions of a business. While various definitions of strategy exist, the following meanings of the term will be used in this book.

Strategy is the art of coordinating the means (money, human resources, and materials) to achieve the ends (profit, customer satisfaction, and company growth) as defined by company policy and objectives. In more pragmatic terms, strategy is defined as *actions* to achieve objectives at three distinct levels:

- *Corporate strategy.* At this level, strategies are developed at the highest echelons of the organization. The aim here is to deploy company resources through a series of actions that would fulfill executives' vision and objectives for the future of the organization.
- *Midlevel strategy.* At this juncture, strategy operates at the business-unit, department, or product-line level. It is more precise than corporate strategy.

Typically, it covers actions covering a three- to five-year period and focuses on fulfilling specific objectives. Also at this level, strategy covers two zones of activity: actions to create and retain customers and actions geared to preventing competitors from dislodging the company from its market position (i.e., seizing market share and customers).

- *Lower level strategy or tactics.* This level requires a time frame shorter than those at the two higher levels. Normally, it correlates with a company's or business unit's business plan and the yearly budgetary process.

In everyday application, tactics are actions designed to achieve short-term objectives while in support of longer term objectives and strategies. Tactics are precise actions that cover such areas as pricing and discounts, advertising media and copy themes, the Internet, sales force deployment and selling aids, supply chain methods and relationships, training, product packaging, value-added services, and the selection of market segments for product launch.

Roots of Strategy

Strategy is not a business term. Its origin comes from the ancient Greeks and is derived from the word *strategia* (or *strategos*), meaning to lead an army or generalship. Generals over the centuries have relied on military strategy to conquer territory and gain power. This meant imposing their will on others and maximizing the impact of their economic and human resources to achieve their goals. The challenges were to outwit competing forces, gain territory and power, and conserve resources while expanding their influence. While the terminology may vary, these challenges are not much different from those of businesses.

Although the destructive aspects of war are not present in business, there is a reasonable parallel when you consider the financial demise of organizations, the layoffs of vast numbers of personnel, and the closing of physical plants, as well as the severe economic impact and societal disruption that damages and, in some instances, decimates regions, cities, and local communities.

Thus, business scholars, executives, and line managers readily accept the comparisons and find purpose in studying the chronicles of documented military strategies that span 2,500 years of recorded history. They serve as an excellent additional resource for business study, as opposed to relying only on a single dimension of examining current business practices. This book, therefore, is organized to blend selected historical lessons with modern business practice to provide a solid platform to understand, develop, and apply competitive business strategies.

As you begin the book, you will find that the preceding points are organized into nine irrefutable strategy rules. Each is supported with actual case examples, quick-tip guidelines, and applications. Although the rules have been arranged to be read in sequence, delve into the book at any place according to your interests or to solve a specific problem.

Organization of the Book

The book consists of nine chapters, which are designated as strategy rules:

Strategy Rule 1. Shift to the Offensive: Turn a Risky Competitive Situation into a Fresh Market Opportunity. When boldness meets caution, boldness wins. Based on historical evidence in the military and other disciplines, this rule points out that standing still—stalled by lack of ideas and immobilized by fear—can fester into severe problems. You will learn how to use a ten-step guideline to customize strategies and stay on the offensive. You will also find out how to focus on five critical areas to estimate your competitive situation.

Strategy Rule 2. Maneuver by Indirect Strategy: Apply Strength against Weakness. Indirect strategy is the invincible rule that stands out as one of the consistently successful ingredients of a business plan. You will learn how to apply strength against a competitor's weakness, resolve customer problems with offerings that outperform those of your competitors, and achieve a psychological advantage by creating an unbalancing effect in the mind of your rival manager.

Strategy Rule 3. Act with Speed: The Essential Component to Secure a Competitive Lead. Rarely has an overlong, dragged out campaign proved successful. Exhaustion through the draining of resources damages more companies than almost any other factor. This rule shows you how to identify the barriers to speed and to maintain ongoing momentum.

Strategy Rule 4. Grow by Concentration: Deploy, Target, Segment. This hard-hitting rule means adopting a strategy that concentrates resources where you can gain superiority in selected areas. Doing so creates the following effect: You emerge stronger than your competitor in key segments of your choosing. This rule also shows you how to integrate concentration into your business plans and strategies.

Strategy Rule 5. Prioritize Competitive Intelligence: The Underpinnings of Business Strategy. This rule shows you how to utilize the tools of competitive intelligence, identify the behavioral personalities of competitors, and select agents to augment traditional competitive intelligence techniques.

Strategy Rule 6. Align Competitive Strategy with Your Corporate Culture: The Lifeline to Your Organization's Future. Corporate culture is the operating system and nerve center of your organization. It guides how your employees think and react when entangled in a variety of hot spots. A supportive corporate culture drives forward-looking business decisions, generates customer loyalty, and ignites employee involvement. You will learn how to identify the characteristics of high-performing business cultures and see their impact on developing competitive strategies. You will also learn techniques to reenergize your company's culture as you reinvent your competitive strategies.

Strategy Rule 7. Develop Leadership Skills: The Moral Fiber Underlying Business Strategy. Leadership is about responsibility and accountability, as well as achieving corporate and business unit objectives. Leaders inspire their people, organize actions, develop strategies, and respond to market and competitive uncertainty with speed and effectiveness. This rule shows you how to personalize your leadership style and power-up your business strategy through effective leadership.

Strategy Rule 8. Create a Morale Advantage: Engage Heart, Mind, and Spirit When All Else Fails. This rule is about the human side of competitive strategy. Success in competitive encounters is an issue of morale, discipline, and trust. In all matters that pertain to an organization, it is the human heart that reigns supreme at the moment of conflict. You will learn how to overcome the barriers to success resulting from poor morale and use techniques to activate high morale.

Strategy Rule 9. Strengthen Your Decision-Making Capabilities: Fortify Intuition, Enhance Business Experience, Expand Knowledge. Studying business history in general and probing past campaigns in particular can sharpen your decision-making and strategy skills. Since no event is a stand-alone occurrence, you will learn how to link one event to another, uncover the roots of a problem, and find out what went right or wrong. Using case examples, you will learn how to analyze market events and interpret competitive encounters despite the fog of uncertainty.

Deliberately and systematically following these nine enduring rules of competitive strategy can help you overcome the obstacles that have crushed other managers. Then, integrating them into your business plans and strategies can increase your chances of triumphing over rivals looking to oust you from the marketplace. On the other hand, deliberately avoiding them or even minimizing their use places you at a distinct competitive disadvantage, whereby your influence could be reduced to a marginal segment of the market.

This book will guide you to think like a strategist and become more proficient as you fight today's competitive battles. It will show you how to prepare yourself, your subordinates, your company, or business unit to win—to win customers, to win market share, to win a long-term profitable position in a marketplace, and to win a competitive encounter before a rival can do excessive harm.

To further assist you, there is a unique feature at the end of each chapter. The *strategy diagnostic tool* provides a reliable performance measure to support you in building, evaluating, and monitoring your business strategies. Since business campaigns succeed or fail in accordance with the rules of strategy, this tool assesses the risks and overall viability of a competitive strategy and helps you gauge the likelihood of its success. Also, there is an outline of a *strategic business plan* in the appendix that can serve as the starting place for developing your strategies, as well as a format for presenting your ideas to management.

The time-tested rules and numerous real-case examples in the following chapters will serve as venerable guiding principles as you get to the roots of competitive strategy. Good luck!

The Author

Norton Paley's consulting and training services have been used by organizations such as American Express, IBM, Detroit Edison, Chrysler, McDonnell-Douglas, Dow Chemical (U.S., Pacific Rim, European, and Canadian operations), W. R. Grace, Cargill (United States, Japan, and United Kingdom), Mississippi Power & Light, Chevron, Ralston-Purina, Johnson & Johnson, U.S. Gypsum, Celanese, Hoechst, and Holderbank (Switzerland), as well as scores of small and midsize companies. His international training experience includes lecture tours in The Republic of China and Mexico (the latter sponsored by the U.S. Information Agency). He has been a seminar leader on marketing management, strategic planning, and competitive strategy programs for over 20 years at the American Management Association.

Mr. Paley's corporate management experience includes managerial positions with McGraw-Hill and John Wiley & Sons. He has published seven books:

How to Develop A Strategic Marketing Plan (translated into Chinese and Turkish). "This book is both intellectual and practical...an interesting vehicle for presenting detailed planning concepts...it is clear and well-organized" (T. J. Belich, in *CHOICE*)

The Manager's Guide to Competitive Marketing Strategies, 3rd edition. "The ultimate weapon in helping a company increase market share" (*Atlanta Journal-Constitution*)

Marketing for the Nonmarketing Executive: An Integrated Management Resource Guide for the 21st Century

Pricing Strategies and Practices

The Marketing Strategy Desktop Guide, 3rd edition

Successful Business Planning: Energizing Your Company's Potential (rights purchased for Russian translation)

Manage to Win (Indian edition published by Viva Books, New Delhi). "A book too forceful to ignore" (*The Financial Daily*)

Additionally, Mr. Paley has developed three comprehensive training modules that include computer simulations for marketing learning systems segmentation, targeting and positioning, and marketing planning systems. His byline columns have appeared in *The Management Review* and *Sales & Marketing Management* magazines.

Strategy Rule 1

Shift to the Offense: Turn a Risky Competitive Situation into a Fresh Market Opportunity

Chapter Objectives

1. Utilize a five-step process when preparing for the offensive.
2. Exploit market opportunities by holding reserves.
3. Assess areas of internal and external friction that prevent shifting to the offensive.
4. Employ the strategy diagnostic tool to assess Rule 1.

Introduction

Throughout business history one indisputable strategy rule has prevailed with steadfast certainty:

> When on the defensive, plan for the offensive.*

---

* The rule also reflects the accumulated experiences from other fields of endeavor, particularly the enduring lessons of over 2,500 years of recorded military history.

This rule can be expressed another way: If you act defensively to protect a market position from an aggressive competitor, take the preliminary step of moving to the offensive. The alternatives—stalled by lack of ideas, immobilized by fear, standing still, or restrained by blurred imagination—can fester into severe problems. In all likelihood, your competitors feel no less anxiety when faced with similar choices: They must react or retreat. Thus begins a mental sparring game with choices vacillating from firm resolve to paralyzing fear. The essential point: If you are entangled in a tough competitive situation, it is in your best interest to develop a proactive posture, rather than to languish in indecision.

Further, as rivals see you initiate actions in an organized and consistent manner, your actions make a striking impression. They signal your determination to shift the psychological advantage in your favor. Therefore, maintain the momentum by preplanning a variety of ready-to-use initiatives, such as launching new or modified products; delivering add-on services; kicking-off preemptive promotions to high-potential segments; or taking advantage of emerging, neglected, or poorly served market segments. Otherwise, by avoiding such planning you fumble between fight-or-flee decisions. Also, if you habitually back away or remain overlong on the defensive, you end up overlooking ripe opportunities when they arise. Just as serious, you send a clear message to competitors that you are not going to challenge any of their aggressive moves.

Equally severe, over the long term, this could place your entire operation at undue risk. "If you don't take a risk on a new idea, that in itself becomes a risk," asserts President Tsuyoshi Kikukawa of **Olympus Corp.** Consequently, planning for the offensive dictates that you sharpen your ability to monitor competitors' actions in such areas as price movements, product innovations, promotion drives, or supply chain initiatives. This also means taking the pulse of the market and observing the dynamics of your customers' buying behavior. Such activities will give you time to organize your plans and shape your strategies.

Consequently, at times you hold back, checking your own inclination to act impulsively, but at other times you give it free rein. Going on the offensive provides forward motion, which in itself takes on the spirit of determination. It is the combination of offensive action and boldness that should define your highest level of accomplishment and underscore your primary challenge: *Develop winning business strategies that ultimately result in winning customers and preserving the growth of the market.*

Further, the higher you rise on the executive ladder, the more likely it is that you will face situations that require you to act with some measure of daring—tempered, that is, by preparing precise estimates of your available resources, assuring reliable competitive intelligence, and employing strategic plans to prevent pointless actions based on guesswork and emotion. In this way, boldness and courage become prudent and responsible acts of your leadership style.

Shift to the Offensive with Boldness

Other factors being somewhat equal, when boldness meets caution, boldness wins. Undue caution is countless times worse than excessive courage. When the history of business is considered, enterprise, accented with determination and purpose, rather than by impulse, leads more often than not to successful performance. Audacity also has a powerful psychological impact on the competition, whereas excessive caution is always handicapped by a loss of stability, initiative, and momentum. Therefore, when in doubt, some action is better than no action. The following case provides a tangible perspective to the rule.

Case Example

Apple Inc. has long enjoyed the role of niche player in its comfortable position within the personal computer (PC) market. In 2002, however, overall PC market sales plateaued with Apple's share still hovering in the three percent range, which it held for several years.

Presented with the uncertain consequences, yet driven by the need to grow or go downhill, Apple's assertive CEO Stephen Jobs made the aggressive leap out of his comfort zone and moved aggressively into the mainstream of the consumer electronics market. Apple launched iPod, a digital handheld music player smaller than a deck of cards. Original industry estimates pegged sales at 12 million units, with sales growth predicted at a rate of 74 percent annually for several years.

What initially gave the iPod a boost was its ability to work with most PCs. As a further advantage, the product was at the introduction stage of its market and product life cycles. The fresh opportunity? By moving fast enough, Apple could establish a viable market position and avoid fighting excessive battles against other aggressive competitors waiting to latch on to the spiraling product category. Therefore, market leadership for such music devices was up for grabs to the company that moved boldly and rapidly, established a brand identity, and continued to maintain a competitive lead. By 2007, Apple again made industry headlines and rankled leading cell-phone makers with its innovative iPhone.

What can you learn from the Apple case? Boldness is a practical course of action if it is supported by sound business plans, reliable estimates of needed resources, ongoing competitive intelligence, and a solid commitment to your company's long-term objectives. Such considerations prevent you from taking mindless risks by entering markets that are indefensible and beyond your company's financial resources, supply chain strengths, and employees' capabilities. Consequently, if you support your plans with sound deliberations, your boldness is justified. Further, you vastly improve your chances of extending your horizons, with the strong likelihood that you will end up with profitable outcomes.

There is, however, the highly realistic situation in which you face a market circumstance that is time sensitive and requires you to take some action or forfeit a potentially lucrative opportunity to a competitor. This means having to live with gaps in competitive intelligence. In such a case, you have little choice but to move ahead with the information on hand, relying on your experience, self-confidence, flexibility, and intuition to face up to the market condition as it unfolds (more on intuition later).

"Sometimes you have to operate on instinct and fire before you have all the information on hand. That can flush the opportunities out of hiding," declares CEO Masamitsu Sakurai of **Ricoh Co.** In all, the impulsive display of boldness is not the central issue. Rather, an ordered and calculated approach reduces the level of risk embedded in most audacious actions. "I would define true courage to be a perfect sensibility to the measure of danger, and a mental willingness to incur it, rather than that insensibility to danger of which I have heard far more than I have seen," said General William T. Sherman.

Therefore, when you are ready to deploy your human, physical, and financial resources to implement your plan or encounter a competitor, do so by using a systematic approach. The following pages provide details about the five preliminary steps you should take before any decisive action. These include:

1. Make reliable estimates and calculations.
2. Use a diagnostic tool such as a SWOT (strengths, weaknesses, opportunities, threats) analysis to assist in clear thinking and to prioritize your actions.
3. Hold reserves to exploit opportunities.
4. Assess levels of creativity and innovation.
5. Evaluate the ability of your personnel to deal with friction originating within the company and from the marketplace.

Make Reliable Estimates and Calculations

What exactly should you estimate? To make meaningful estimates, focus on five categories: *internal and external relationships, leadership, seasonal forces, market selection,* and *policy.*

Internal and External Relationships

Look at the interpersonal relationships that exist among your business units or product lines. Determine if individuals with whom you work are motivated to act with some measure of enthusiasm and élan or are sluggish, bored, and fearful of personal risk. Are employees managed with respect, fairness, and honesty? Is there a corporate culture that unites them in purpose, staying power, confidence, and a sincere willingness to overcome the inevitable competitive obstacles? Is there a working environment that encourages creativity and the forward thinking that results in innovative products and services? The following case examples illustrate these points:

Case Example

Vice-Chairman Robert Lutz of **General Motors (GM)** had to deal with difficult internal situations when he arrived on the job with the troubled company. With their originality and flair, GM's concept cars could have turned consumer heads, but they never appeared as production models.

Lutz found a corporate culture focused more on engineering processes than on generating excitement among consumers. Job number one was to dissolve the adversarial and communications gaps among engineers, auto designers, and financial personnel. This meant convincing "bean counters" to invest in styling, persuading the engineering and design heads to stop fighting each other, and trashing bureaucratic rules that prevented the fresh designs from reaching the dealers' showrooms.

To heighten enthusiasm in the ranks, Lutz forced design studios to compete in "sketch-offs." To foster internal and external relationships, he tapped GM's global resources and encouraged personnel to share ideas and develop models more quickly and cheaply. Formerly, global units had tended to act autonomously, duplicating each other's work no matter how much money GM wasted.

In your estimates, do not overlook the caliber of leadership, which is often pivotal to the outcome of a marketing effort and a competitive encounter. "Leadership is influencing people by providing purpose, direction, and motivation while operating to accomplish the mission and improve the organization."* Other definitions of leadership refer to various qualities of a manager:

- An *insightful* manager recognizes from the onset the eventual impact of new competitive entries, changing customers' behavior, industry restructuring, and environmental influences. Once this has been noted, an incisive manager is able to act rapidly and boldly to create opportunities and defuse threats.
- A *straightforward* manager's employees will have no doubt about how and when rewards and reprimands would be handed out.
- A *compassionate* manager respects people, appreciates their industry and toil, and empathizes with them under adverse situations.

---

* Source: *U.S. Army Leadership Field Manual.*

- A *bold* manager gains market advantage by seizing opportunities without hesitation.
- A *strict* manager is demanding and dedicated to the long-term objectives of the organization. In turn, personnel are disciplined because they are respectful of all those strong-minded attributes, yet fearful of reprimands—or worse.

Case Example

Siebel Systems Inc., the corporate maker of software to manage salespeople and call centers, endured a difficult period between 2000 and 2004. Managers watched the likes of SAP, Oracle, and salesforce.com pounce on their company and take away customers. Revenues during this period tumbled as a result of the fierce competition. A new CEO, George T. Shaheen, was brought in to lead a turnaround.

Central to the evolving strategy is innovative software, which breaks the core customer relationship management (CRM) product into LEGO-like components that can run on standard **Microsoft** and **IBM** platforms. The product's built-in uniqueness means that customers can buy the pieces they want and snap them together with other software, rather than buying complete features-laden, complex software systems.

Asked if he was under pressure to produce quick results, Shaheen replied, "I think leadership is sort of simple. To be a good leader, you have to have a vision. You have to have courage to go after it. And you have to have a reasonable track record of success. At the end of the day, you have to do what you think is right."

Seasonal Forces

This estimate focuses on the consequences of natural, climate-related conditions. It influences how you manage your business within the variables of weather and logistics, such as seasonal outcomes of winter's cold and summer's heat. For instance, what impact does weather have on industries such as home building and road construction, on transporting materials to meet critical delivery schedules, on installing communications systems, and on supplying energy, as well as on all the ancillary products and services associated with these industries? Further, what is the weather or seasonal impact on food supplies, fashion, entertainment, and retailing?

Market Selection

With this estimate, your concerns focus on the efficient movement of your products and services throughout the supply chain to targeted markets. Also, your estimates of the physical and economic conditions of a territory or market can help immeasurably in determining the ease or difficulty of serving various segments that may be a city or a continent away. Much anxiety about conducting even the most complex business transactions over extended distances has dissolved with the global use of the Internet and advances in technology. In its place, attention has shifted to defining markets by closing cultural and behavioral distances.

For instance, consider the boom in Latin America over the past decade. This vast region experienced explosive growth resulting from a variety of dramatic changes, such as lowering of trade barriers, declining inflation and steadier prices, economic recovery and rising incomes (in some countries), increasing democratic freedom, and the profusion of dazzling new communications technologies. Influenced by these dynamic changes, enterprising managers discovered attractive opportunities by segmenting markets and then targeting the identified groups.

In another example of precise market selection, **Hyundai Motor Co.** pushed into the hottest emerging markets against more established rivals. In India, it reached a strong number two position and is vying for top spot in small cars. In China, where Hyundai began selling cars in earnest only in 2003, it skyrocketed to the top spot in 2005, and it also leaped into a leading brand position in Russia. Hyundai's strategy is to select growth markets and outflank competitors by being first to take over smallish segments. Then, once a foothold is established, the company pushes out to additional segments, supported by modern plants with economies of scale to operate profitably.

Policy

This estimate covers the foundation operating guidelines that control an organization or a business unit. Policy forms a tangible imprint of your company's ethical and operating procedures. It gives your organization consistency and a distinctive personality. While policy making may be outside your area of input, it does heavily influence your ability to:

- Select the types of strategies to grow your business
- Determine the parameters by which you can innovate and compete for market advantage
- Attract new talent and assign others to new levels of authority and responsibility
- Secure your position on the supply chain, with particular attention to solidifying relationships and deterring competitive threats
- Deploy your financial and human resources to exploit market opportunities

Policy, therefore, holds a legitimate and powerful grasp on your business plans. It impacts your ability to cultivate the natural growth of the market. It controls

Exhibit 1.1 SWOT Analysis

Strengths	Look objectively at your organization's or group's unique strengths, not just its physical resources. Identify the special skills inherent in your organization that would permit you to push the boundaries of innovation and discovery. Also, single out any unique characteristics in corporate culture, leadership, internal communications, products, systems, technologies, and processes.
Weaknesses	Determine what weaknesses you see among the preceding factors. Look at possible bottlenecks that could prevent implementing business plans. Examine what can be revamped, reorganized, or discarded. Estimate at what costs in time, money, and resources drawbacks can be remedied.
Opportunities	In scanning the customer, competitor, industry, and environmental situations, what opportunities do you see for your business? What openings exist to displace the competition, expand the company's entry into new markets, serve new customer groups, or generate new revenue sources? How would you define the opportunities for long-term growth versus short-term limited payouts?
Threats	What immediate and longer term threats do you anticipate and how are you going to face them? Are advances in technology outpacing your company's financial and human capabilities to keep up? What impact will new global competitors have on your company's ability to maintain a profitable market presence? What governmental, environmental, or legislative issues are looming to hinder your ability to operate profitably?

the strategy options you can use to avoid hostile price wars and similar damaging activities, especially against competitors with little interest in nurturing the long-term prosperity of the marketplace.

Prioritize Your Actions

How you prioritize your actions depends on how you assess your company's or group's strengths, weaknesses, opportunities, and threats. Known as the familiar SWOT* analysis (Exhibit 1.1), this is a widely used and time-tested approach, especially within the framework of shifting to the offensive. When employed in a group setting, it provides a highly reliable technique for estimating your situation from internal and external vantage points.

---

* A SWOT analysis is a reliable management tool. There are others readily available. If another tool provides a suitable assessment, use it.

Exhibit 1.2 The Interface of SWOT and Business Planning

After conducting a SWOT analysis, incorporate the results into your business plan. Doing so permits you to develop explicit objectives and strategies to reinforce your company's *strengths*, eliminate or reduce its *weaknesses*, exploit *opportunities*, and lessen the impact of *threats*. This means that you add greater precision to your plans and your ability to:

- Improve communications with individuals who may be widely dispersed or when it is necessary to spell out the reasons for selecting overly challenging objectives.
- Deploy personnel so that the skilled and the not-so-skilled work in harmony at the critical junctures where speed and precision are needed to take advantage of a sudden opportunity or to downgrade a problematic weakness.
- Monitor the behavior of personnel so that the more aggressive individuals continue to act boldly but not impulsively and the more reticent ones act with some bravado, yet do not retreat prematurely under competitive pressure.
- Shape bold strategies to shift to the offensive.

Even with your best efforts at conducting an accurate SWOT analysis, the reality is that your business plans can come apart when your original assumptions about market conditions do not materialize. Unforeseen situations, such as unexpected price wars, disgruntled channel members along the supply chain, changing industry priorities, or shifting demographics in your primary segments, can loom as potential threats. Also, threats from overly aggressive competitors can hamper your best efforts.

Therefore, build enough flexibility and what-if scenarios into your business plans. Also, develop second-tier objectives in the event that the primary ones are no longer within reach. Then you can react in sufficient time to remedy situations that might otherwise deteriorate beyond a reasonable chance of recovery. Additionally, consistent with the strategy of shifting to the offensive, you gain the advantage of selecting the time and place to activate your (revised) plans. For instance, you can choose to enhance your product's competitive position with new applications or you can launch into emerging market niches to open new revenue streams (see Exhibit 1.2).

As for additional possibilities, you can push pricing decisions down to field personnel, initiate fresh promotional incentives, establish joint ventures to access new technologies, and install innovative value-added programs within the supply chain. All such actions would enhance your managerial competence in applying bold strategies. Again, this is part of the overall framework of "when on the defensive, plan for the offensive."

An unacceptable alternative is to "stand still, immobilized by fear" as you see your marketing efforts falter and risk giving your competitors open access to your markets and supply chain. Then, you may have to endure the disagreeable task of going back to management hoping for more resources—with the strong odds of being turned down. An additional and workable solution to permit flexibility is to retain sufficient reserves. This maximizes your ability to maintain an elastic position against known and unknown circumstances, as well as to exploit fresh opportunities.

Hold Reserves to Exploit Opportunities

The purpose of holding reserves is to ready yourself for unexpected opportunities, such as repositioning some resources to take advantage of sudden openings in previously neglected or poorly served market segments. Additionally, you open up options to react confidently when faced with a sudden threat from a competitor.

Reserves apply to a variety of choices. You can hold in reserve a portion of your operating budget to exploit the rapid adoption of your product by funding additional advertising, offering extended warranties, or adding customer-service personnel. Also, there are options in packaging innovations; value-added services; additional buying incentives; pricing discounts; or new promotional formats, such as streaming messages over the Internet that could be timed to coincide with customer acceptance or to upset unexpected competitive moves.

Further, you can organize your personnel to form a "flying" reserve, which gives you the flexibility to shift individuals from one location to another, depending on evolving market conditions. Not allowing for a reserve is a serious omission in your business plan. You forfeit options to follow up immediately and take advantage of budding opportunities. Exercising those options, however, is only relevant if you know how and when to activate your reserves. This requires reliable market intelligence, with particular emphasis on understanding your competitors' strategies, assessing the qualities of their personnel, determining their strengths and weaknesses, and evaluating the overall caliber of their leadership.*

Therefore, the end purpose for holding reserves is totally compatible with the strategic rationale of what you are in business to accomplish:

- Satisfy the needs and wants of your customers in a manner that is superior to that of your competitor.
- Maintain the long-term growth and viability of your markets.

Consequently, the piercing signal for you is *not* to commit all your resources at one time. Being able to provide an energetic response to market and competitive opportunities at a time and place of your choosing places you in a superior position to turn a doubtful situation into a decisive victory.

---

* The following quote from a former time and field captures the idea: "Pay attention to your enemies, for they are the first to discover your mistakes."—Antisthenes

What follows is a fundamental principle and a central theme of the rule "shift to the offensive": Never remain completely passive, even when under a disadvantage from a larger company. Simply keep moving. Stay on the offensive, no matter what the effort. Just do not lose the momentum! The following case illustrates a practical application for holding reserves.

Case Example

Arrow Electronics is a wholesaler of computer chips, capacitors, and myriad other components that make subsystems for PCs, cell phones, and autos. The problem is that some customers, such as the huge contract manufacturers like **Flextronics** and **Solectron**, buy more parts directly from suppliers and thereby cut out the middleman.

Arrow recognized that it excelled in one area that represents a huge reserve, as well as a core competency, not available from direct suppliers: the ability to provide superior services. Certainly, services are not unique. But they can be an exceptional counterstrategy against contract manufacturers that usually rely on middlemen to provide a variety of operational services associated with sound customer relationship management.

For Arrow, services include generous financial arrangements, on-site inventory management, parts-tracking software, and chip programming. The company also provides software that helps customers identify parts that are easily available, soon to be obsolete, or that can be made according to new environmental standards.

In one instance, Arrow voluntarily kept three material planners at a customer's location to handle parts flow. They were also alert to looking for ways to substitute parts that Arrow could supply for less. The result was that total procurement costs for Arrow's customer dropped 20 percent and new products got to market in just seven months versus the previous fifteen months.

In sum, reserves allow you to shift to the offensive against competitors when and where needed. You thereby gain the advantage of choosing one point in your competitor's weaknesses on which to concentrate your efforts.

Quick Tips: What to Look for in a Competitor

- How entrenched is the competitor in the marketplace (with special attention to the supply chain, relationships with key customers, and the stage of customers in the buying cycle)?

- What image does the competitor have among intermediaries and customers? How does it compare with yours?
- Is there a strategy pattern the competitor uses to launch a new product, maintain its market position, or respond to a pricing attack? How would you describe its strategies: aggressive, moderate, or minimal?
- How committed is the competitor's management to sustaining an ongoing effort in the market: heavy, moderate, or light commitment?
- How interested is the competitor's senior management in the ongoing development of the market?
- What is the overall quality of the competitor's personnel in terms of skills, training, discipline, and morale? How would you rank its leadership?

Assess Levels of Creativity and Innovation

At the turn of the twenty-first century, companies were thought to be secure if they were connected to the knowledge economy as a competitive advantage. However, that, too, has become commoditized and moved, in part, to China, India, Hungary, Russia, and the Czech Republic. What has evolved for many companies is a shift from the knowledge economy to the creativity economy. The movement has powerful implications for sustaining the offense and developing it into a core competence.

This is not accomplished just by maintaining an edge in technology; the movement applies creativity, imagination, and innovation to any functional areas of the business that could impact offensive strategy. Nor does the movement appear to be a short-term fad. Increasingly, companies are embracing the change. Universities are introducing numerous programs in creativity and a growing number of design labs are catering to companies hungering for dynamic paths for growth. The movement aims at creating new consumer experiences, not just modifying existing products or introducing line extensions.

At the forefront of the movement are **Procter & Gamble** and **General Electric**, followed by Apple, **3M**, Microsoft, **Sony**, **Dell**, IBM, **Wal-Mart**, and scores of others. These companies are driving forward armed with an intimate understanding of consumer behavior. They are honing their ability to determine what consumers want even before people can articulate them.

A process has evolved to foster creativity. It consists of the following steps:

- *Maintain ongoing observation.* Go beyond the conventional market research studies and get into the streets, talk with customers, and observe their buying habits. Cultural anthropologists call the process an ethnographic study. In the business world, for instance, managers at **Gap Inc.** observed that social shopping in pairs or threesomes is the norm for women shoppers in its stores. Noting the consistency of such behavior, Gap management enlarged dressing rooms to accommodate that buying pattern.
- *Create models, videos, or simulations.* Using a hands-on, interactive experience permits concepts to come alive. The feedback helps designers decide what to modify or discard. It thereby reduces the risk of failure and quickens the product launch.
- *Develop a narrative.* Designers have found that wrapping a potential new product around an emotional story connects with consumers and improves the chances of success. The design of a new line of watches and driving shoes, for example, captured the story of the Mini Cooper's cool urban driving experience—a happening related to the driver, not the car.
- *Install a process.* The object is to make the creative process an ongoing and intrinsic function of the organization. That means understanding the culture of the organization and undertaking changes where appropriate.* Sometimes the changes can be wrenching experiences, but the potential payout is enormous. As one business consultant termed it, "The creativity economy is more anthropology, less technology."

The following case illustrates these points.

Case Example

Motorola Inc. floundered for several years and let the landmark product it invented, the cell phone, slip from its grasp as it watched **Nokia** seize the lead. Under new leadership, a bold strategy emerged for new products that pushed the company into an upturn. The turnaround strategy is cleverly defined by Motorola's design chief: "Design leads, engineering follows. Now we base everything on some experience we want to project and then have the engineering team help us get there."

This approach is a 180° reversal of Motorola's former approach to product development, which was characterized by an operating culture where engineers created the product innards and designers encased it in plastic or some other material. Equally important is the cultural changeover. Rather than develop one startling product, a process and a mindset are currently in place to crank out one hit after another. The emphasis is now deeply anchored to creativity and innovation.

---

* See Strategy Rule 6: Align Competitive Strategy with Your Corporate Culture.

Results for 2005–2006 indicated that the new strategy was paying off. At a time when rivals **Nokia** and **Samsung Electronics** reported lackluster results, the once troubled **Motorola** gathered speed.

While creativity, imagination, and innovation apply most often to new products and services, you can also activate the process and add originality to your business strategies and give them a unique character. You thereby ensure that you are elevating your thinking to a new dimension and not simply repeating the off-the-shelf actions of a past period.

Evaluate the Ability of Personnel to Deal with Friction

As you shift to the offensive, you must reckon with the reality that damaging market forces can create *friction*. In turn, the likelihood is that your business plans will be delayed, modified, or totally trashed. The eminent business scholar Michael Porter* describes his famous five competitive forces, which are characterized here as potential types of friction:

1. Rivalry among existing firms in an industry
2. Threat of potential new entrants
3. Bargaining power of buyers
4. Threat of substitute products or services
5. Bargaining power of suppliers

Friction takes other inhibiting forms. You will find its damaging effects in the psychological sphere when exhibited by employees through:

- Low morale
- Fear and uncertainty
- Lack of trust resulting from ineffectual leadership
- Depleted levels of energy due to negative perceptions about unfolding market events
- Discouragement and even defeatism resulting from aggressive competitive actions

Friction also springs from apathy and customers' buying resistance to your product offerings or indifference to a new incentive program by key players in the supply chain.

Also, there is deeply rooted friction from inexperienced or poorly trained employees. They are the ones who are not up to the rigors of implementing offensive strategies, which require discipline, cooperation, commitment, and mental agility

---

* See Porter, M. *Competitive Strategy, Techniques for Analyzing Industries and Competitors*. (New York: Free Press, 1980).

to stay balanced against the gyrating ups and downs of competitive encounters. Consider, too, the internal friction from the organizational logjams and layers of management that prevent the clear communication of directions from senior management to field personnel. Also, due to an inability to obtain correct data in a useable format, friction surfaces when decision-making managers are unable to estimate the situation correctly and act rapidly on an opportunity or a competitive threat. Still other areas of friction come from the internal staffs that fail to provide timely financial, legal, logistical, and other vital information. All these represent points of friction.

The result is that friction continues its insidious damage by fostering errors that slow down day-to-day operation. The causes of friction are limitless. Thus, you should be fully aware of the ones that surround you and then do what is necessary to limit the irreparable damage to your overall business plan as you shift to the offensive.

Limit the Negative Effects of Friction

You have a built-in "X" factor—an immensely powerful internal mechanism that can override some of the damaging effects of friction: *intuition.** You are able to experience intuitive assistance in a variety of ways, such as by instinct, insight, hunch, or "gut" feeling. You also receive impressions in the form of a vision (clairvoyance), hearing (clairaudience), and sensation (clairsentience).

To some, intuition suggests an ethereal quality that cannot be pinned down when it comes to developing actionable strategies and reducing the dire effects of friction. Yet, there is sufficient empirical evidence and pragmatic experience to indicate that you can confidently rely on this innate quality to take action and avoid being immobilized by friction. This is especially so where rational thinking and market intelligence do not produce trustworthy solutions. Therefore, in a competitive confrontation, what ultimately plays out is a contest of your mind's creativity and originality against that of a competitor's mind. Consequently, when you need to rely on intuition, you can engage the mind and free it in a purposeful direction. Intuition is personal and takes on your inborn personality as your mind goes to work on a problem.

Strategy, then, is a blend of art and science that embodies the distinct imprint of the individual, made distinguishable by the infallible quality of intuition. Therein lies the genius of managers who overcome friction and rise to success. For some strong-willed, in-charge managers, nothing can replace intuition. Even when managers lack originality and a determined personality and even with the support of advisors and subordinates, there are decisive moments when they must take counsel

---

* Intuition is discussed in greater detail in Rule 9: Strengthen Your Decision-Making Capabilities.

within themselves, make decisions, and move forward. Accordingly, trusting in intuition to contend with friction is a reliable technique. Seasoned managers consciously know the value of intuition in emergencies. Yet, they are also fully aware that intuition must be anchored to solid experience, judgment, and ongoing training—as well as to the rules suggested in this and the following chapters.

Proof that your intuition is working appears when you reach a comfort level where relatively sound decisions come almost automatically so that you know intuitively, for instance, that one strategy is more likely to work, whereas another will not. Notwithstanding the voluminous quantities of knowledge to support decision making, savvy managers understand that most market events are more or less hidden in a mist of uncertainty. Uncertainty is further magnified as they recognize that competing managers must rely on intuition as well. Moreover, rival managers are also surrounded by dynamic physical and psychological forces that create damaging friction and cloud the competitive scene.

Although your weightiest decisions are often made on uncertain premises, it would be totally false to assume that success is a matter of sheer luck. It is not luck in the ordinary sense that brings achievement. In the long run, so-called chance favors the skilled and intuitive manager.

In sum, taking all actions that permit you to shift to the offensive is one of the most productive and winning rules of strategy. Therefore, even when forced to the defensive, as in protecting your share of the market, your best course of action is to plan for the offensive. The next rule, maneuver by indirect strategy, sets the tone for your offensive movements.

Before moving on, review Rule 1 using the strategy diagnostic tool to assess how this rule would affect your strategy.

* * *

Strategy Diagnostic Tool

Strategy Rule 1: Shift to the Offensive

Part 1: Indications That Strategy Rule 1 Functions Effectively (Contributes to Implementing a Successful Competitive Strategy)

1. Managers display a proactive, shift-to-the-offensive mindset in competitive situations.

    ☐ Frequently   ☐ Occasionally   ☐ Rarely

2. Senior management permits risk taking with no serious repercussions for negative results.

    ☐ Frequently   ☐ Occasionally   ☐ Rarely

Shift to the Offense ■ 17

3. Midlevel managers are adept at making timely maneuvers to block rivals from doing excessive harm to the company's competitive position.

   ☐ Frequently   ☐ Occasionally   ☐ Rarely

4. Managers act with boldness, which has a positive psychological impact on employee behavior and morale.

   ☐ Frequently   ☐ Occasionally   ☐ Rarely

5. The organizational design permits personnel to communicate effectively and move rapidly on time-sensitive market opportunities.

   ☐ Frequently   ☐ Occasionally   ☐ Rarely

6. Managers recognize that employees' experiences, flexibility, self-confidence, and ability to use their intuition are reliable attributes for reaching trustworthy decisions.

   ☐ Frequently   ☐ Occasionally   ☐ Rarely

Part 2: Symptoms That Strategy Rule 1 Is Functioning Ineffectively (Detrimental to Implementing a Successful Competitive Strategy)

1. Managers and staff are stalled by complacency or apprehension and lack fresh initiatives for growing the business.

   ☐ Frequently   ☐ Occasionally   ☐ Rarely

2. Personnel are overly preoccupied in defending an existing market position, with negligible time and effort engaged in searching for new market and product opportunities.

   ☐ Frequently   ☐ Occasionally   ☐ Rarely

3. Midlevel managers and staff are stalled by fear of what competitors might do, which prevents developing a vigorous response strategy.

   ☐ Frequently   ☐ Occasionally   ☐ Rarely

4. Senior management exhibits unwarranted caution, which prevents them from acting on new market prospects.

   ☐ Frequently   ☐ Occasionally   ☐ Rarely

5. The organization tends to lose its momentum.

   ☐ Frequently   ☐ Occasionally   ☐ Rarely

The ratings for Parts 1 and 2 are qualitative assessments of managers' overall ability to shift to the offensive and execute an effective competitive strategy. Based on a diagnosis of your company's situation, use the following remedies to implement corrective actions:

- Attack complacency: Organize ready-to-implement contingency plans to react promptly to market opportunities or competitive threats.
- Prepare for new opportunities: Use market and competitive intelligence to support decisions and reduce the risks inherent in making bold moves.
- Instill self-confidence: Initiate specialized training related to enhancing job skills and improving discipline, as well as elevating employees' self-assurance, flexibility, and use of intuition.
- Regain momentum: Use market segmentation analysis to understand how and where to deploy resources and thereby prevent taking mindless risks in indefensible markets.
- Encourage an entrepreneurial mindset: Require key personnel to present proposals that would take the company into new markets, products, and services with the potential of establishing additional revenue streams. (Use the strategic business plan in the appendix to this book.)
- Support a market-driven orientation: Create cross-functional strategy teams that encourage creativity and visionary thinking.
- Prepare for unexpected events: Hold reserves.

Remedies and actions: _____

_____

Finally, to demonstrate the far ranging impact of this rule, the following company problems linked to the offensive come from a survey of chief executives of medium and large organizations (company names withheld to maintain confidentiality):

"Facing an avalanche of competition and bogged down by erratic marketing."
"My company is losing momentum and losing its way."
"Need to find the way back to the market where we got creamed."
"We need to energize the product line plagued by missed opportunities, sluggish sales, and dismal profits."

Strategy Rule 2

Maneuver by Indirect Strategy: Apply Strength against Weakness

Chapter Objectives

1. Distinguish between direct and indirect strategies.
2. Identify the components of an indirect strategy.
3. Use a six-step process to develop an indirect strategy.
4. Integrate indirect strategies into your business plan.
5. Employ the strategy diagnostic tool to assess Rule 2.

Introduction

All that precedes and follows any entry into a new market involves committing substantial financial, human, and material resources. It means supporting and training personnel, designing and producing products, providing backup services, determining pricing, activating the supply chain, and launching promotions. All is gambled on winning or losing in a blinding market moment—notwithstanding that the final outcome may not be apparent at the instant, but may take months or even years to play out. Nonetheless, the imprint has been set in such a moment as the organization demonstrates its inner capabilities and justifies its future existence as a viable entity.

With so much at stake, is there some overriding strategy rule, any point of commonality, that envelops the numerous market variables and competitive obstacles? Is there one prevailing rule that is more likely to succeed, which ultimately affects whether customers adopt or decline a product; determines if a product line remains viable; and, long-term, decides if the organization grows and prospers, languishes as an also-ran, or totally fails? Such a universal rule is embedded in the power of the *indirect strategy*.

The Power of Indirect Strategy

Seemingly unimposing by name, the indirect strategy endures as the indomitable rule that consistently towers as one of the vital ingredients of a business plan. The indirect strategy operates on three dimensions. First, the strategy is anchored to a line of action whereby you apply your strength against a competitor's weakness. The essence of the move is to maneuver so that your rival cannot, will not, or simply lacks the capability to challenge your efforts. Second, concurrent with activating indirect moves against a competitor, you focus your attention on serving customers' needs or resolving their problems in a manner that outperforms your competitors' actions. Third, you aim to achieve a psychological advantage by creating an unbalancing effect in the mind of the rival manager. That is, by means of distractions and false moves you make it appear that you are launching your effort directly at the competitor's strengths. However, your true purpose is to target his vulnerabilities.

Additionally, the psychological effects of the indirect strategy seek to disorient the competing manager, causing him or her to waste time, effort, and resources in the wrong direction or, expressed differently, to make costly and irreversible mistakes. All three applications serve the strategic purpose of reducing any resistance leveled against your efforts. In turn, you can then utilize the full power of your resources without wasting them on strength-draining actions. Look at the indirect strategy as an encounter of manager against opposing manager: your experience and skill pitted against those of your opponent.

Quick Tips: Indirect Strategies in Action

1. Bypass traditional channels: offer online and inexpensive MBA degrees to midlevel managers in their 30s and 40s, thereby circumventing traditional university venues and focusing on an emerging market segment (United States).

2. Deliver a differentiated product: produce a cassette with the advanced digital capabilities of a compact disc (United Kingdom).

3. Set a market-driven price: distribute a cell phone inexpensive enough for people in remote rural villages (China and India).
4. Use targeted promotions: offer flavored mineral water aimed at upper-class consumers (Germany).
5. Create a unique supply chain: provide refills of ink cartridges left at coffee shops; dispense video cassettes at commuter train stations (United Kingdom).

Examples abound over the past decades to document the advantages and applications of the indirect strategy:

- German and Japanese automakers first entered the North American automobile market with small cars during the energy crisis. This market essentially had been neglected by domestic manufacturers during the 1970s and poorly served during the 1980s and into the 1990s. Once embedded, they expanded into full lines of cars covering all price segments of the market with resounding success.
- With the introduction of its Miller Lite product in 1975, **Miller Brewery** identified the light-beer category—now the largest segment of beer drinks—as an emerging market.
- **Dell Computer** bypassed the traditional distributor channel through retailers and other intermediaries. Instead, it sold directly to the end user with a build-to-order strategy that complemented its low-price approach.
- **Apple Computer** became a dominant factor in schools early on, specifically serving that segment, left unserved by **IBM**, with computer hardware and software.
- **Wal-Mart** originally opened its stores in towns with populations under 15,000 that were totally ignored at that time by the leading retailers.

With the abundance of historical business examples, it is now safe to conclude that there is never any justification for you, or any manager, to undertake a direct frontal attack by applying your strength against an opponent's strength in today's competitive market. It is wasteful in time and resources, violates the primary principle of strategy, and rarely achieves its goals. Further, a direct strategy means confronting a stronger competitor head on where there is little or no differentiation in product features, quality, performance, and service and where there is no identifiable advantage in price, promotion, distribution, technology, leadership, or caliber of personnel.

Under these circumstances, losses to a company that forces a direct confrontation are enormous in terms of human, financial, and material resources. That is, a company that moves directly against a competitor actively defending a market position can exhaust itself before reaching its sales, market share, and profitability goals. Even if a company achieves some minor objective through a direct effort, such as scoring minimal sales or nominal market share, meager resources will remain to move forward and secure enough market share to reach profitable levels.

Developing an Indirect Strategy

There is a logical and systematic process you can use to develop an indirect strategy. The process consists of six steps:

1. Define your strategic goals.
2. Determine the resources needed to achieve the strategic goals.
3. Gather competitive intelligence.
4. Establish security.
5. Implement the indirect strategy.
6. Develop a poststrategy.

Define the Strategic Goals

In broad terms, usually covered by a three- to five-year planning period, describe the strategic goals or direction of your organization, business unit, or product line. Then specify measurable objectives. Also, be certain that your objectives complement your organization's overall mission so that you do not go running off in ill-advised directions.

For instance, **Motorola** strived to maintain momentum after relinquishing market leadership to **Nokia**. Most notably, Motorola launched its Razr phone, which caught Nokia and other competitors by surprise and captured the attention of a worldwide audience. Razr came about when Motorola executives decided to buck the growing industry trend to load up phones with cameras and stereo speakers that made them heavier and bulkier. Instead, Motorola introduced a half-inch thick phone with sleek lines and a shimmering keypad. The Razr did more than just ring up sales; it achieved the strategic goal of getting customers around the world looking again at Motorola as "creative, cool, and sexy."

To assist in developing your strategic goals, begin by answering the following questions.

What Are Our Organization's Distinctive Areas of Expertise?

Here is where you look at your organization's (or business unit's) competencies, not only in physical resources but also in leadership and managerial capabilities. Specifically, you want to look at the following:

- Competitive strengths of your product or service based on customer satisfaction, profitability, and market share
- Depth of your relationships with distributors and end-use customers
- Efficiency of existing production capabilities
- Level of morale and discipline of your in-house personnel and field sales force
- Availability of sufficient finances to carry on operations during good and bad times
- Commitment to new product development and use of current technologies
- Quality of customer or technical services

What Should Be Our Organization's Business during the Next Three to Five Years?

Here is where you pinpoint the market segments or categories of customers you are likely to serve. For instance, note any standout trends that would connect your strategic goals to customers' needs and wants. Doing so projects your thinking into new product development, instead of relying primarily on the longevity of existing products to sustain company growth. By taking the long-term view, you begin to position your business strategically for the future. In turn, that view determines the breadth of existing and new product lines, which helps you identify new market opportunities. If you are too narrow in defining your strategic goals, the resulting product and market mix will be generally narrow and possibly too confining for growth.

On the other hand, defining your business too broadly can result in spreading capital, people, and other resources beyond the capabilities of your organization. Therefore, look to create a comfortable balance by positioning your business somewhere between the extremes of the two positions.

What Segments or Categories of Customers Will We Serve?

Customers exist at various levels in the supply chain and in different segments of the market. At the end of the chain are end-use consumers with whom you may or may not come into direct contact. Other customers along the chain serve as intermediaries and typically perform several functions. They include distributors who take possession of the products and often serve as a warehousing facility. Still other intermediaries repackage products and maintain inventory-control systems to serve the next level of distribution. Value-added resellers provide customer service, technical advice, computer software, or educational programs to differentiate their products from those of competitors.

Examining the existing and future needs at each level of distribution helps you project the types of customers you want to target for the three- to five-year period covered by your strategic goals. Similarly, you will want to review various segments and target those that will provide the best opportunities over the planning period.

What Additional Functions Are We Likely to Fulfill for Customers as We See the Market Evolve?

As competitive intensity increases worldwide, each intermediary customer along the supply chain is increasingly pressured to maintain a competitive advantage. This question asks you to determine what functions or capabilities are needed to solve customers' problems. More precisely, you are looking beyond your immediate customer and reaching out further along the distribution chain to identify those functions that would solve your customers' *customers'* problems. Such functions might include providing computerized inventory control, after-sales technical support, quality control programs, just-in-time delivery, or financial assistance.

What New Technologies Will Our Firm Require to Satisfy Future Customer Needs?

Look again at the previous question and think about the practices of your industry. Examine the impact of technologies to satisfy your customers' needs. Look at where your company ranks with the various technologies and types of software used for product design and productivity, manufacturing, and distribution systems. Look, too, at the continuing changes in information technology and business intelligence and the resulting effects on product innovation and market competitiveness. Also appraise such emerging technologies as expert diagnostic systems, dashboards, and other business performance management (BPM) systems for problem solving. Look at the rapidly changing communications systems to manage and protect an increasingly wireless enterprise.

What Changes Taking Place in Markets, Consumer Behavior, Competition, Environment, Culture, and the Economy Will Impact Our Company?

This form of external analysis permits you to sensitize yourself to critical issues relating to your business from which you can develop an indirect strategy. To further illustrate strategic goals and the application of the indirect strategy, the following case shows how an organization with only a minimal presence in a product line engaged two market leaders firmly entrenched in a specialized market.

Case Example

Wal-Mart Stores decided to attack two of the largest consumer-electronics chains, **Best Buy** and **Circuit City**. Wal-Mart began by sprucing up the interiors of many of its electronics departments and adding several high-end products, from Sony liquid-crystal-display televisions and Toshiba laptops to Apple iPods.

However, simply adding products to its line was not a strong enough move to unseat the two leaders. To win with an indirect strategy Wal-Mart would have to locate and strike them at an area of greatest vulnerability. The company located such a decisive area of weakness and struck.

The vulnerability lay in the most profitable line of business for Best Buy and Circuit City: extended warranties. For these two companies, real earnings were not in the sale of the electronic gadgets. Rather, they were in the sale of multiyear protection plans on TVs, computers, and other items that were aggressively hawked by the retailers' salespeople. For Best Buy, warranty sales accounted for more than a third of its operating profit; they accounted for all of Circuit City's.

Wal-Mart pulled out all stops and concentrated on that line of attack by launching extended warranties on TVs and computers at prices that averaged 50 percent below the two retailers. "Profit on extended warranties has always been the Achilles' heel of Best Buy and Circuit City," declared an industry analyst.

What was Wal-Mart's strategic goal? The company's push into consumer electronics is part of its long-term objective to attract more upscale shoppers. Wal-Mart managers observed that wealthier consumers shopped mostly for food and cleaning products. To get them to go through the whole store, managers reasoned that upgrading the electronics departments as well as other high-end product lines would achieve that strategic aim.

Strategic aim, therefore, is the first step in developing an indirect strategy. The purpose, again, is to guide your activities with discipline, rather than wander off in several directions, expending resources without a defined purpose and a measurable end. There is a second but no less important purpose to the strategic aim: to coordinate with the overall corporate direction. This broader view provides credibility when seeking approval for a budget from senior management.

Quick Tips: Developing an Indirect Strategy

1. Find an unattended, poorly served, or emerging market segment.
2. Create a competitive advantage by using the marketing mix (product, price, promotion, and distribution) in a configuration that cannot be easily matched by competitors. That is, use your strength against the weaknesses of your opponent.

3. Mobilize all available resources to fulfill the unmet needs and wants of the market in a strength-conserving manner and thereby solidify relationships with your customers.
4. Expand into additional segments of the market with new or modified products and services.

Determine the Resources Needed to Achieve the Strategic Goals

This step refers to the quantities and types of resources you will need to get the job done. Whereas financial and material resources are the raw essentials that are budgeted according to the long- and short-term objectives, the focus here is in deploying people resources that ultimately decide the fate of any campaign.

Deploying and leading people effectively means harnessing their energies without exhausting them. Positioning human resources also requires the shrewd management of assets in two ways: using the *normal* and the *extraordinary*. The *normal* relates to all those activities you would normally use to market your products and satisfy customers' needs. In contrast, the *extraordinary* points to those unique—and indirect—strategies and innovative offerings that can significantly outstrip those of your rivals. Usually associated with differentiated products, value-added services, and other components of the marketing mix, they are difficult for even the most aggressive competitors to imitate, at least for the short term. (See Motorola and Wal-Mart examples, which tie in with indirect strategies.)

Extraordinary examples can cover a variety of categories, as long as they fit the definition and criteria of an indirect strategy. Exhibit 2.1 offers a source of ideas from which to develop an indirect strategy. To activate an indirect strategy, engage with the *normal* initially to match the competitor's offerings. Then, swiftly launch with the *extraordinary* to wrest the advantage from competitors.

The essential point is that you cannot achieve any measure of success without devising strategies that artfully coordinate both the ordinary and extraordinary forces. Otherwise, the business suffers the consequences of inching along in a direct, laborious, and resource-draining manner that can only end up with marginal or failed performance.

As part of resources, there is yet another significant issue that decides the outcome of a campaign: *leadership*.* This factor is characterized by the following attributes:

---

* See greater detail in Strategy Rule 7: Develop Leadership Skills.

Exhibit 2.1 Sources for Creating Indirect Strategies with the Extraordinary

Product or Service	Price	Marketing	Supply Chain

Quality
Features
Options
Style
Brand name
Packaging
Sizes
Support services
Warranties
Returns
Versatility
Uniqueness
Utility
Reliability
Durability
Patent protection
Guarantees
	Discounts
	Allowances
	Payment period
	Credit terms
	Special financing
		Advertising:
		  Print
		  Broadcast
		  World Wide Web
		Personal selling:
		  Incentives
		  Sales aids
		  Samples
		  Training
		Sales promotion:
		  Demonstrations
		  Contests
		  Premiums
		  Coupons
		  Manuals
		Telemarketing
		Internet
		Publicity
			Channels:
			  Direct sales force
			  Distribution
			  Dealers
			Market coverage:
			  Warehouse locations and proximity to customers
			Inventory control systems and ordering
			Physical transport

- A leader should be able to recognize changing competitive, customer, industry, and environmental circumstances and act boldly to create opportunities and diffuse threats.
- A leader should be consistent so that employees will have no doubt about how and when rewards and reprimands are meted out.
- A leader should demonstrate the humanistic trait showing respect and appreciation for individuals' hard work and loyalty, and empathize with them under adverse situations.
- A leader must have courage to seize opportunities without getting mired in indecision.
- A leader should be demanding and dedicated to the objectives of the organization. It is also important that he or she show enough discipline and avoid displaying any weakness that will discourage personnel and create doubt.
- A leader should value the psychological facets of competitive strategy and use it to manage people by reaching their hearts and minds. Equally important is understanding the significance of achieving a psychological advantage over a rival manager, as illustrated in the following Heublein example.

Gather Competitive Intelligence

Central to employing an indirect strategy is the reliability of competitive intelligence because there is no meaningful way that you can determine what constitutes *direct* or *indirect* if you do not know the direction in which you are positioned against your competitor. This point is illustrated in the following classic example:

Case Example

Heublein, the producer of Smirnoff vodka, enjoyed a leading brand position with a dominant market share for two decades. At one point Smirnoff was attacked on price by a competing brand, **Wolfschmidt**, which was then produced by The Seagram Company Ltd.

Wolfschmidt employed a strategy of pricing its product at $1.00 a bottle less than the price of Smirnoff and claimed the same quality. Recognizing a real danger of customers switching to Wolfschmidt, Heublein needed a creative strategy to protect its market dominance. Managers examined a number of options:

Lower the price of Smirnoff by $1.00 or less to hold on to its market share.

Maintain the price of Smirnoff but increase advertising and promotion expenditures.

Maintain the price of Smirnoff and hope that current advertising and promotion would preserve the existing Smirnoff image and market share.

While some options were attractive, they were all obvious and mainly direct approaches. Instead, Heublein decided on an indirect strategy. First, it raised the price of Smirnoff by $1.00 and thereby positioned its flagship product to preserve the premier image, market position, and brand identity it already enjoyed. Next, Heublein introduced a new brand, Relska, and positioned it head to head as a fighting brand against Wolfschmidt's price and market segment. Using that product entry as a means of diverting the opposing manager's attention from other actions, Heublein introduced still another brand, Popov, at $1.00 less than Wolfschmidt.

That action had the decisive effect of enveloping Wolfschmidt by using the normal and the extraordinary. The result was that, during the 1980s, Smirnoff remained number one in cases of all imported and domestic vodkas shipped in North America, with Popov in the number two position.

The Heublein case clearly demonstrates how strategy lies in the sphere of the indirect strategy. It also indicates the indispensable need for competitive intelligence to provide data on how the opponent's product is positioned—in this case, the relationship of Smirnoff's position to Wolfschmidt's—and to show strategy options to create an indirect strategy. For instance, the physical act of repositioning Smirnoff upscale and then introducing the threatening fighting brand, Relska, directly at Wolfschmidt temporarily distracted the rival manager into inaction.

That move also demonstrates the psychological effect of the indirect strategy on the Wolfschmidt managers. They were sidetracked and unbalanced by the threat to their market share. By indirectly dislocating Wolfschmidt, the strategy reduced their capabilities to resist. Further, the total envelopment created by the positioning of the three products caused a psychological paralysis that reduced any further action by Wolfschmidt.

The essential reason for drawing on competitive intelligence is to develop competitive strategies. Use it to unravel behavioral, transactional, and historical data about your markets, customers, and competitors. You accomplish this with the increasing availability and sophistication of data mining software programs.

Surfacing in the mid-1990s, data mining evolved to become a vital component in shaping most business strategies. In recent years, there have been additional breakthroughs in obtaining business intelligence with the vast improvements in analytical tools built around the Internet. Data mining is a computer-based process that uses a variety of analytics to discover patterns and relationships in data that may be used to make valid predictions. Initially, data are extracted from a company's internal data warehouse and placed into a data-mining database.

Then, the multifaceted technology of the Internet is used to acquire finite intelligence about groups' and individuals' shopping patterns. As individuals surf the Web to make inquiries and do their shopping, they leave valuable information about their transactions. These hidden files or tags called "cookies" are deposited on their computers. Software programs then use those files to track and analyze online behavior. Such data become the foundation to design a product or service offering built around a one-on-one approach. In general, this process is not a do-it-yourself project. Numerous vendors with the appropriate software are available to install the system in a company or provide for outsourcing.

In the everyday application of competitive intelligence, you can benefit by actively tuning in to market conditions, following buying patterns, and interpreting competitors' moves, thereby providing meaningful options on how to maneuver resources. To trigger your thinking and assist in devising an indirect strategy, here are several applications of intelligence:

- **American Express** gathers existing customer information from its call centers and uses the data to make highly targeted cross-sell and up-sell offers to customers.
- **Charles Schwab** compiles routine requests from its investor accounts to form a comprehensive profile of its customers. The in-depth data are then reconfigured and applied to such revenue and profitability goals as customer

retention, cross-selling, and up-selling. The aim is to maintain an advanced level of differentiation in an industry that has become intensely competitive.
- **Kodak** collects, analyzes, and acts on precise data to constantly monitor customers' buying behavior, from which it personalizes messages with product offerings.
- **Hilton Hotels** probes for guest information from its hotels and resorts and makes it available in useable form to hotel managers over the Internet. Those managers can then develop new or improve existing services for their guests, manage corporate loyalty programs, and manage marketing campaigns.
- **Verizon** leverages its data to develop profiles for each of its telecommunications customers. Then, based on each customer's history and preferences, it offers products and services that are likely to appeal to each individual.

In all of these applications, managers used the hard data to form a two-directional indirect strategy. First, they determined the buying patterns within their respective markets and used the information to devise products, services, and positioning strategies. Second, they combined the customer data with the intelligence acquired about their competitor's product and service offerings, along with their overall marketing tactics. From that total intelligence, managers shaped and fine-tuned their own indirect strategies.

Establish Security

In step three, make certain that your market positions are secure. Security is particularly vital as you concentrate your resources on a specific segment of the market. A dramatic example of such a breach in security is the case of **Xerox**, which still serves as a constructive lesson for today's challenging competitive conditions.

During the 1970s, the company concentrated almost entirely on selling its large copiers to big companies. Whether by choice or by management's myopic view of the total marketplace, it was blind to protecting its dominant market position. In so doing, Xerox managers left a huge gap for enterprising Japanese copier makers to enter an unattended market with small copiers aimed at the huge number of small and midsize companies. Once established, rivals such as **Canon** and **Ricoh** took the natural route by expanding their lines of copiers, which finally encroached on Xerox's hold of the large-company segment. The subsequent precipitous drop in Xerox's market share has taken decades to recover. Meanwhile, the intruders embedded themselves solidly in the North American market.

Consequently, it is in your best interest to maintain ongoing intelligence about those competitors that might make an indirect attack against you. The attack could be an unguarded gap in your product line, a flawed service capability, a poorly served user segment, or an overlooked geographic area. The broader aspects of competitive intelligence procedures were reviewed previously. However, do not overlook using the eyes and ears of those involved in one-on-one contact with local customers and competitors, as well as those with functional and tactical responsibilities—for instance:

- *Sales reps* should observe local activities about new promotional incentives by competitors, product line, and territory. They should even provide input about their counterparts' morale and levels of selling skills.
- A *sales manager* can look from a higher vantage point to observe how competitors' salespeople are deployed, types of sales aids used, forms of communication, and methods of compensation. The sales manager would also look at the patterns of customer movements within a defined territory, such as changes in sales and product usage from previous periods.
- *Advertising* and *market research* people can interpret a variety of signs, such as how a new competitor deviated from its usual mix of media and promotional methods, details about impending acquisitions or divestitures, new product developments that might signal an entry into your segment, and any significant shifts in competitors' key managerial positions.
- *Financial* or *accounting* personnel should examine the financial health of a competitor by looking at such indicators as debt-equity structure, inventory turnover, capital resources, credit rating, and any other clues that would highlight strengths and vulnerabilities.
- *Production* personnel would look for signs that reveal production processes, plant locations and logistics, age of equipment or technologies, quality control systems, and similar areas from which an accurate assessment can be made.

Specifically, here is what you want to find out to strengthen security:

- What is the competitor's strategy for expanding out of its present market segments? Does it have sufficient manpower resources? Would it place your market position in jeopardy?
- What is the competitor's financial picture? Can it hold out in a prolonged competitive campaign?
- What new products, technologies, or services are under development that would constitute a threat?
- How is the competitor organized and staffed? What is its staff's level of morale, discipline, training, and overall preparedness? Are there weaknesses you can exploit?
- What is the caliber of the competitor's leadership? Are they skilled and experienced? Would their behavior tend to be bold, timid, or passive if challenged?
- Where are the competitor's potential areas of vulnerability by product depth, product quality, service, price, distribution, and reputation? (Also review Exhibit 2.1 to conduct an in-depth comparison.)

Implement the Strategy

Your purpose in implementing an indirect strategy is to win: To win customers, to win market share, to win a strong presence on the supply chain, or to win whatever other objectives you have set to achieve—and to do so profitably. Above all, the strategy to win is not meant to exhaust resources through competitive confrontations resulting in slugging it out with competitors over a prolonged period of time. The following case describes the implementation of an indirect strategy.

Case Example

Caterpillar Inc. operates a remanufacturing plant that works on giant 12-cylinder worn-out truck engines. Workers dismantle them and rebuild them for renewed lives of hauling or pushing huge loads. It is hardly glamorous work to blast away at the buildup of concrete-hard coatings of grease, dirt, and oil to reach the thousands of parts before rebuilding the entire engine.

The company got into the business indirectly. It dates back to a favor Caterpillar reluctantly did for its customer, **Ford Motor Co.** To lower its own costs, Ford's truck-making subsidiary wanted a source to produce rebuilt engines, which generally sell for half the price of new ones. To maintain relationships as a supplier to Ford for new engines, Caterpillar consented and opened a repair shop to accommodate its customer.

Organizationally, the remanufacturing operation is separate from Caterpillar's core business of producing new truck engines and land-moving equipment. It has since become one of the most profitable divisions of Caterpillar and is slated to grow by 20 percent a year for the foreseeable future.

Where does implementing the indirect strategy come in? First, the remanufacturing unit listened and responded to a customer-driven request to provide a service. Doing so incurred out-of-pocket expenditures to build a plant to do the low-tech, dirty work of dismantling and rebuilding truck engines. However, it was justified by the long-term strategic aim of maintaining a profitable customer relationship.

Second, Caterpillar's entry into the engine rebuilding business circumvented a recurring problem that exists in its mainline business: having to deal in the new equipment part of an industry that is highly cyclical and subject to wide swings resulting from economic conditions. In contrast, revenues and profits from remanufacturing services continue to climb with some degree of stability. Estimates show that the segment has excellent long-term growth potential.

Third, Caterpillar management realized that this ancillary business represented a hidden and indirect opportunity to enter another market segment: an underserved and needy segment consisting of a sizeable number of users that could not afford new equipment. Fourth, selling rebuilt products at discount prices formed a solid indirect strategy that blocked makers of knockoff parts out of the lucrative aftermarket and permitted Caterpillar to profit again and again from the same parts.

Develop a Poststrategy

Industries, markets, and products live within the inevitable cycles that span introduction, growth, maturity, and decline. These life cycles are often rocky and can result in the demise of a company or product if not given serious attention. Therefore, it is worth your time and effort to develop a poststrategy to manage these unavoidable and volatile life cycles—on your terms.

Without a poststrategy, your overall business plan is incomplete and most often will lead to shoot-from-the-hip actions. Further, it places you at the mercy of market and competitive forces with no orderly strategy to extricate yourself from the market or the ability to turn the situation to your advantage. With a poststrategy linked to your business plan, you are better able to anticipate potential areas of vulnerability and mark out possible actions. For instance, you can implement contingency plans to cope with the sudden intrusion of competitors arriving with new technologies, counter their claims of above-average product quality and performance, or respond to their threat to push you out of the market.

A poststrategy exists on two levels. First, due to aggressive competitors making it untenable for you to remain in the market, you can suddenly exit the market after fulfilling obligations to workers, customers, and communities. On the other hand, you may choose to reduce your presence in a market by removing products using a deliberate phased withdrawal (see Exhibit 2.2). Second, you can opt to remain in the market for the long haul. That choice entails enhancing customer relationships, prolonging the life cycle of an existing product, replacing failing products with new offerings, and managing other parts of the marketing mix (e.g., promotion, communications, supply chain, or pricing).

Some companies have been especially successful in prolonging product life cycles. One noteworthy example is **3M**'s Scotch Brand Tape. The product has gone through several cycles and has become an accepted fixture in industrial and consumer markets spanning several decades. The tape has been modified and reformulated for heavy factory use, and it has been repacked to suit a variety of applications in office and consumer markets.

The focus here is on the second level: a poststrategy for enhancing relationships with customers, prolonging the life of a product or service, and extending the long-term growth of the market. Providing, that is, you believe there is sustainable life in the market and your product and that your presence fits the long-term goals of your organization or business unit.

Exhibit 2.2   Criteria for a Poststrategy to Phase Out a Product

Provide answers to the following questions:

What is the market potential for our product? What is the product's contribution to our company's bottom line?

What are the chances of our product being displaced by another product or technology?

What competitive advantage might we gain by adding value, modifying the product, or creating other differentiating features and benefits?

What would we gain by repositioning our product against the competitor's comparable product to create an indirect strategy?

How many resources (materials, equipment, people, and money) would be available by eliminating the product?

How good are the opportunities to redeploy resources to a new product, market, or business?

What value does the product have in supporting the sale of our other product lines? Is the product line filled out sufficiently to prevent our customers from shopping elsewhere?

Is the product useful in blocking a point of entry against competitors? (See Xerox example.)

What impact will removing the product have on executives' time, the deployment of the sales force, and dealer and customer relationships?

Prolonging the Life of Existing Products and Creating New Product Opportunities

The following five product categories represent approaches you can take to extend the life cycle of existing products, as well as identify new product opportunities. Three of those categories permit you to do something physical to the product, while the remaining two maintain the same product characteristics. This approach is rooted to the generally accepted definition that a product is new when it is *perceived* as new by the customer, regardless of anything physical you do to the product.

1. *Modification*: Add or alter product features to conform to customer requirements, as well as differentiate your offering from that of your competitor. You retain the same number of product lines and products.
2. *Line extension:* Add more variety to your offerings, while keeping the same number of product lines. Doing so results in a higher number of products to accommodate for additional product applications. The purpose is to segment the market and offer more dedicated choices.

3. *Diversification*: Enter a new business with a modified product line and new applications. The advantage is that you spread the risk and capitalize on fresh opportunities and a new revenue stream; providing the move complements your organization's long-term objectives, and there are sufficient resources to enter new markets.
4. *Re-merchandising*: Create new impressions through an innovative marketing effort. Use the same products to the same markets. The intent is to generate excitement and stimulate sales.
5. *Market extension:* Enter new markets with existing products and repackaging (where applicable). The purpose is to expand your market base into additional niches that have evolved or have been previously underserved.

Enhancing and Prolonging Customer Relationships

All your efforts to develop indirect strategies, confront competitors, and maintain a presence in the market should converge on a single bottom-line outcome: to create and keep a customer. Use the following guidelines to assist your efforts:

- *Explore methods to reduce customers' returns and complaints.* Begin reviewing product quality, order fulfillment procedures, and postsales services. Doing so may entail working directly with customers to get to the source of the problems. Improvement in those areas would likely cut down customers' administrative overhead.
- *Look at processes to speed up production and provide value-added services to benefit your customers.* These two separate functions may need joint work teams to explore processes for redundancy and ways to smooth out procedures. The importance of doing so would cut costs and thereby strengthen relationships with your customers.
- *Work with customers to improve their market position and image.* This means examining your customers' marketing efforts and providing consultative services. Likened to consultative selling, your investment in this effort will not only improve relations, but will likely generate additional sales. Also, this approach can apply to other areas of the business, such as offering procedures that would lower customers' production costs.
- *Determine the value of adding a name brand to the line and assess its impact on your customers' revenues.* A name brand has sales-building potential and the advantages of exclusivity. The one consideration, however, is whether management is sufficiently committed to sustaining the value and integrity of the brand over the long term.
- *Explore which product or service benefits would enhance your customers' operations.* You have a wide range of tangible and intangible possibilities open to you. It would be best to team with customers and lean heavily on their needs (see the previous Caterpillar case example).

- *Examine if improving reordering procedures would impact revenues.* The time it takes filling and shipping orders and, subsequently, the speed of reordering impact the bottom-line. Examine the available automated logistics and just-in-time delivery systems touted by such companies as **FedEx**, **DHL**, and others. Such systems should slash customers' delivery costs.
- *Work with customers to extend the product's life.* Explore ways to promote more frequent usage of the product, locate new users for the product, find more uses for the product, and develop new uses for the product's basic materials. Accomplishing these tasks may include the acquisition of products, licensing of technology, or pursuing other types of joint ventures—potentially creating a competitive edge for you and your customers.

Summary

As noted throughout this chapter, you should avoid a direct strategy in favor of an indirect strategy. The object of the indirect strategy is to circumvent the strong points of competitive resistance. Doing so permits you to concentrate in those markets of opportunity where you can achieve a competitive advantage. The end result is that you conserve resources by employing a strength-saving approach and improve the chances of achieving your objectives—all of which translates into profitable growth.

Three primary characteristics distinguish the indirect strategy. First, your aim is always to apply strength against your competitor's weakness. Further, your actions should be implemented with speed so that a rival cannot, will not, or simply lacks the capability to challenge your efforts. Second, concurrent with triggering an indirect strategy, your intention is to serve customers' needs or resolve their problems with offerings that outperform those of your competitors. Third, by using speed, concentration, and surprise, you look to achieve a psychological advantage by creating an unbalancing effect in the mind of the rival manager. The purpose is to reduce any competitive resistance leveled against you, thereby allowing you to use the full potential of your resources.

Before moving on, review Rule 2 using the strategy diagnostic tool to assess how this rule would affect your strategy.

* * *

Strategy Diagnostic Tool

Strategy Rule 2: Maneuver by Indirect Strategy

Part 1: Indications That Strategy Rule 2 Functions Effectively (Contributes to Implementing a Successful Competitive Strategy)

1. Managers intentionally integrate indirect approaches into their business plans, thereby increasing the success rate of their strategies.

    ☐ Frequently   ☐ Occasionally   ☐ Rarely

2. Managers understand that acquiring skills for implementing indirect strategies opens the mind to ideas that challenge market leaders, even when limited resources are available.

☐ Frequently ☐ Occasionally ☐ Rarely

3. By actively probing for the competitor's vulnerabilities, managers are better able to allocate resources into differentiated products and value-added services that outperform those of the competitor.

☐ Frequently ☐ Occasionally ☐ Rarely

4. Managers deliberately work at exploiting the psychological benefits of unbalancing and distracting the competing manager into making false moves and costly mistakes.

☐ Frequently ☐ Occasionally ☐ Rarely

5. Managers intentionally select strategies that reduce the chances of getting entangled in direct confrontations with competitors, which would result in the unnecessary draining of resources.

☐ Frequently ☐ Occasionally ☐ Rarely

Part 2: Symptoms That Strategy Rule 2 Is Functioning Ineffectively (Detrimental to Implementing a Successful Competitive Strategy)

1. Managers do not attempt to use an indirect strategy to deliver differentiated products that outperform those of competitors.

☐ Frequently ☐ Occasionally ☐ Rarely

2. Managers do not probe for unserved market niches, where there is minimal resistance from competitors and where possibilities exist to establish a foothold and expand into a mainstream market.

☐ Frequently ☐ Occasionally ☐ Rarely

3. Personnel do not rely on competitor intelligence to formulate an indirect strategy.

☐ Frequently ☐ Occasionally ☐ Rarely

4. Managers do not have a benchmarking system to evaluate strengths, weaknesses, or best practices periodically, which can be used to develop competitive strategies.

☐ Frequently ☐ Occasionally ☐ Rarely

5. Personnel do not actively search for new channels to bypass an overcrowded supply chain.

☐ Frequently ☐ Occasionally ☐ Rarely

The ratings for Parts 1 and 2 are qualitative assessments of managers' overall ability to use indirect strategies to execute an effective business plan. Based on a diagnosis of your company's situation, use the following remedies to implement corrective action:

- Use a SWOT analysis, or comparable analytical tool, to help you determine the type of indirect strategies to employ.
- Focus on your current competitive situation and compare it to the objectives you want to achieve. Doing so allows you to gauge if your indirect strategies will produce the desired results.
- Use all available sources of intelligence to interpret your market position, which would provide additional clues to the makeup of your indirect strategies.
- Find an unattended, poorly served, or emerging market segment as a prime target to implement an indirect strategy for market expansion.
- Look to the traditional marketing components of product, price, promotion, and supply chain for ideas on how to configure a strength-against-weakness mix that cannot be easily matched by competitors.

Remedies and actions: _____

_____

Finally, to demonstrate the far ranging impact of this rule, the following company problems linked to the use of indirect strategy come from a survey of chief executives of medium and large organizations (company names withheld to maintain confidentiality):

"Need to outmaneuver local and regional rivals and try to muscle into new markets."

"Need to keep the company from becoming an also-ran in the industry."

"We must find a way to move closer to the end user and develop a true competitive edge."

"We must fend off competition not only from low-cost providers, but also from competitors with more flexible products."

Strategy Rule 3

Act with Speed: The Essential Component to Secure a Competitive Lead

Chapter Objectives

1. Describe the six strategic values of speed.
2. Identify and overcome the barriers to speed.
3. Employ the strategy diagnostic tool to assess Rule 3.

Introduction

Embedded in the rule of speed is a steadfast truism that affects virtually all companies and individuals, which states:

- For companies: Few cases of overlong, dragged out business campaigns have been successful. Exhaustion—the excessive draining of resources—damages more companies than almost any other factor.
- For individuals: Drawn out efforts divert interest, diminish enthusiasm, and depress morale. Individuals become bored and their skills lose sharpness. The gaps of time created through lack of positive results give competitors a greater chance to react and blunt the best intended efforts.

Now, link the preceding statements to the central theme of Strategy Rule 1: Shift to the Offensive, which states that if you act defensively to protect a market position from an aggressive competitor, take the preliminary step of moving to the offensive. The alternatives—stalled by lack of ideas, immobilized by fear, standing still, or restrained by blurred imagination—can fester into severe problems.

Therefore, if you are stuck in an intractable competitive situation, it is in your best interest to develop a proactive posture, rather than languish in indecision. Speed, then, is the unifying element to cement these two ideas and give them wholeness. Yet even with the most convincing evidence and far reaching experiences from such diverse fields as politics, the military, and sports, speed is still largely ignored by many managers.

Strategic Value of Speed

To give pragmatic reality to the rule of speed, let us dig deeper into actual marketplace situations to view the strategic value of speed through the following six propositions. Doing so should elevate this rule to your conscious level in thinking, planning, and implementing.

1. *Timing affects market share, product position, and customer relationships—all of which are difficult and costly to recover once given up.* **Siebel Systems**,* the maker of software to manage salespeople and call centers, saw its revenue nosedive at one point as fierce competition from aggressive rivals blasted in with unrelenting speed and established strong market positions. "They [Siebel] were basically on a clock. If they thought next year would come along without shareholders seeking alternative board members, they were out of their minds," observed one industry analyst. Siebel was indeed on a clock. Management saw few alternatives other than to move as rapidly as possible to develop a response strategy before time ran out. Recognizing that competitors would also react rapidly, Siebel intensified its product development efforts and created a product advantage that it used as the centerpiece for a turnaround. Known as Nexus, the flexible system broke the customer relationship management (CRM) product into components to be pieced together with other software and tailored to customers' applications.

The strategic point: Recovering lost market share, competitive position, and customer loyalty are often more costly, time consuming, and riskier than moving swiftly at the initial signs of threat. The damage is less severe and the odds better for revival.

2. *When a company stalls and loses momentum, it signals an alert competitor to move in and fill a void.* **Motorola Inc.**, the maker of mobile phones, found itself languishing at one point. Only through the energetic efforts of a new CEO did

---

* Siebel Systems is used as a case example in Rule 1 to illustrate the impact of leadership on strategy. In turn, leadership is instrumental in implementing speed.

the company gain enough thrust to move ahead before its chief rivals, **Nokia** and **Samsung Electronics**, could gain market dominance. Similar to Siebel, the centerpiece of Motorola's strategy was a rapid push for new product designs. That move entailed more than just instructing designers and engineers to work out details on a new project. It meant wholesale revamping of the organization's structure, as well as the cultural underpinnings, to accelerate the process. It also meant intrusive, in-your-face leadership by an incessant CEO who monitored progress day by day and hour by hour. Here, too, the clock was ticking for Motorola with continuous and unrelenting certainty.

The strategic point: There are always alert competitors out there probing for weaknesses in a rival company or searching for a poorly served market segment they can exploit. Try not to give it to them.

3. *Speed is a factor in preventing a product from becoming a commodity and possibly causing irreparable damage to a company's reputation.* **SanDisk Corp.**, the maker of the ubiquitous flash memory devices for cell phones, digital cameras, music players, and handheld game consoles, feared that its product would deteriorate into an indistinguishable commodity. SanDisk was enjoying a commanding lead with its product line. Revenues surged an average of 70 percent over a three-year period. However, the business was cyclical and market growth began to plateau with existing product designs. Should competitors get serious about grabbing share from SanDisk by using deep-down pricing, then the market would spin into a profit-draining price war. That painful scenario actually played out in 2005, when flash memory prices plunged and SanDisk's stock dropped by 30 percent in four days.

SanDisk responded with as much swiftness as it could muster. Certainly management had no intention of sitting and waiting for calamity to hit. To preempt additional attacks from competitors, SanDisk moved with all-out speed to adopt a strategy similar to that used by **Intel**: create a brand strategy, similar to the "Intel Inside" approach. But a brand strategy is more than just hyping the name in advertising. The company needed new products. What followed was a hot-footed effort to pump out new lines, including waterproof memory cards, titanium cards, even memory cards that work only if the rightful owner presses a fingerprint on an embedded reader. Company managers also moved quickly to solidify customer relationships by developing dedicated products for such key accounts as **Verizon Wireless** and **Sprint Corp.**, as well as other wireless carriers. Here, too, the inevitability exists that a product moving up the curve of a product life cycle will eventually reach maturity. Competitors will plow in with faster, cheaper, or smaller products and commoditize the market. In many cases, you can delay the dire outcomes through continuous improvement and thereby forestall plateauing for long periods of time.

Strategic point: Product life cycles are a reality that certainly has to figure into your business plan. The appropriate time to develop product and market strategies to forestall maturity and sustain advances is during the growth stage of the cycle.

4. *The risk of losing a viable position in the supply chain occurs by not moving quickly to secure key customers or middlemen.* **EMC Corp.**, the maker of storage software and hardware, made a remarkable recovery after the earlier tech bust. It slashed prices, expanded its product line, and repaired relationships with customers. Through acquisition, licensing, and internal development, EMC moved forward with its strategic objective to help "manage the corporate computing universe." With CEO Joseph M. Tucci at the helm, the company maintained the momentum by rolling out a series of innovative products and by establishing new partnerships. To expand rapidly into new markets and secure a foothold before rivals could respond, Tucci lined up **Dell Inc.** to distribute EMC machines and give it fast access to key markets and customers. Sealing that relationship would secure quicker product adoptions for EMC. "Customers want to buy from fewer suppliers. We want to be one of the companies they depend on for their enterprise solutions and deny competitors entry," declared Tucci.

Strategic point: Timely and efficient distribution is the bedrock requirement of a successful marketing effort. Losing the position within a well-organized supply chain creates a break that is sure to be filled by a rival firm.

5. *Speed adds vitality to a company's operations and behaves as a catalyst for growth.* As a major factor toward competitiveness, it impacts virtually every part of the organization. **Cisco Systems** and **Procter & Gamble** secured vital alliances with important partners early on in China and India. Doing so locked up market positions before their competitors realized what was happening. With enormous growth forecast in those countries for decades to come, the two companies forged superb positions to reap the benefits from the world's highest growth regions. In a similar scenario, **Hyundai Motor Co.** trounced rivals by driving rapidly into emerging markets. In India, it became a strong number two competitor as it pushed for leadership in small car sales. In China, it skyrocketed to the top spot in 2005. The overall strategy was to move quickly into emerging markets before larger competitors could expand. It is part of Hyundai's strategic goal to become the world's fifth largest automaker.

Strategic point: Momentum elevates employee morale and tends to energize an entire organization. Certainly, that is a condition any manager would relish.

6. *A product strategy that integrates speed with technology and efficiency is in the best position to secure a competitive lead.* **Wipro Ltd.**, a software developer and provider of clients' back-office operations, saw the competitive value of speed and accuracy in handling everything from running accounting operations to processing mortgage applications. In that time-sensitive business, the Bangalore, India, company pioneered low prices and dependability. However, management assumed correctly that competitors would soon encroach on its growth path. With tasks that were viewed as overly labor intensive, Wipro managers searched for technologies and systems that would automate many of the manual tasks.

Looking at automaker **Toyota**, Wipro managers became entranced with the superb automation and assembly-line techniques that power its efficiency. Adopting numerous Toyota techniques, Wipro reworked many of its business processes with simple, fast, and smooth techniques that replicate the level of assembly-line proficiency achieved by the automaker. Day and night, thousands of young men and women would take their places at long rows of tables modeled on an assembly line. As in a Toyota factory, electronic displays mounted on the walls would shift from green to red to signal any glitch in the workflow.

Technology also goes beyond the assembly line: Wipro automated processes to skip manual steps and use analytical software to mine data about its clients' customers. That effort is part of the company's all-embracing move to combine speed, efficiency, and technology improvement within a framework of continuous learning and constant change. It is also part of Wipro's overall strategic objective to become the Toyota of business services.

Strategic point: Where cost control, quality improvement, and customer satisfaction are the ongoing trio for measuring successful performance, it is in your best interest to utilize technology to those ends before your competitor preempts you.

Quick Tips: Speed's Guiding Principles

- A sluggish corporate culture is one of the toughest barriers to speed.
- Drawn-out efforts divert interest, diminish enthusiasm, and depress morale.
- Speed is a factor in preventing a product from becoming a commodity.
- Speed is a catalyst for growth.
- Speed does not mean carelessness, recklessness, or shoot-from-the-hip reactions.
- Act boldly, with speed, despite the inevitable uncertainties of customer and competitor behavior.
- Discipline, capability, and skill form the underpinnings of speed.
- Indispensable to speed and success is an organizational culture that is totally customer driven.

Barriers to Speed

Admittedly, there are stumbling blocks that discourage managers from acting with speed. Some are real; others are perceived. Regardless of source, they become genuine barriers to speed. For example, consider the following (mostly legitimate) reasons:

- Lack of reliable market intelligence
- Mediocre leadership that stifles timely and significant progress
- A manager's low self-esteem and indecisiveness as deeply rooted personality traits
- Lack of courage to go on the offensive, triggered by a manager's innate fear of failure
- No trust by manager in their employees' discipline, capabilities, or skills, with a resulting loss of confidence about how they will react in a tough competitive situation
- No trust by employees in their manager's ability to make correct decisions
- Inadequate support from senior management
- Disagreement and open confrontations among line managers about objectives, priorities, and strategies
- A highly conservative and plodding corporate culture that places a drag on speed
- Lack of urgency in developing new products to deal with short product life cycles
- Organizational layers, long chains of command, and cumbersome committees that prolong deliberation and foster procrastination
- Aggressive competitors that strike fear among employees and damage morale, resulting in lost momentum
- Complacency or arrogance as a prevailing mindset

This listing reflects serious barriers to speed, which can solidify as rock-solid deterrents to implementing a business plan with any reasonable chance of success.

At this point, it should be strongly emphasized that speed in no way means carelessness, recklessness, or shoot-from-the-hip reactions. Good estimates, reliable intelligence, and prudent planning are all required procedures. Once these are completed, however, speed is the ally to implementation. Let us examine the barriers in more detail and look for clues you can use to break through the obstacles:

Lack of Reliable Market Intelligence

Even if you conduct formal market research, and even if you have a workable business intelligence system, you should still get actively involved in doing your own grassroots, hands-on competitive intelligence. You will then be more informed and better prepared to move quickly when presented with an opportunity. Here is what to look for:

- Observe changes in the character of your markets. Define your customers by demographic and behavioral characteristics. For instance, look for any unmet customer needs that would enable you to respond rapidly in the form of products, services, methods of delivery, credit terms, or technical assistance.

- Maintain an ongoing dialogue with your customers to find out their most troublesome problems and frustrations. Meet with salespeople. Travel with them. Draw them out on what they see happening in their respective markets. Gain from their insights. It makes you and them more focused and able to react.
- Watch for competitors' alternative and substitute products that could replace your products or services. Examine customer usage patterns. Also observe deviations in regional and seasonal buying patterns. Check for changes from past purchase and usage practices that could translate into opportunities.
- Innovations often occur in selling, especially with the pervasive use of the Internet. Tune in to current trends in promotional allowances, selling tactics, trade discounts, rebates, point-of-purchase opportunities, or seasonal and holiday requirements. Here, again, stay close to salespeople for such information.
- Examine your supply chain and look for opportunities to customize services consistent with the characteristics of the segment. Pay attention to warehousing (if applicable) and what could be fertile possibilities to innovate, such as electronic ordering and computerized inventory control systems that link to data mining capabilities.
- Search for innovations and product-line extensions to maintain an ongoing presence in your existing markets or to gain a foothold in an emerging segment. As illustrated by **Wipro**, harness new technologies that lead to cost-effective and efficient operations. Doing so would make your company more competitive and result in broadening your customer base.

Mediocre Leadership Stifles Timely and Significant Progress

The manager who has little understanding of the market and lacks know-how about the types of actions to take in a competitive situation cannot be an effective leader. Therefore, effective leadership requires that you work with the best information available and act boldly, with speed—even where information about your competitor is faulty or sketchy, and despite uncertainties.* Within that veil of uncertainty, work hard at developing two indispensable qualities to support your decisions. First, even in moments of apprehension, rely on intuitive guidance to find the proper path. Second, elevate your courage and determination to follow your instincts, however faint.†

---

* General Colin Powell indicates that 60 percent of the available information should be sufficient when a decision is required and action needed.
† See discussion on the high value of intuition in Strategy Rule 9: Strengthen Your Decision-Making Capabilities.

A Manager's Low Self-Esteem and Indecisiveness as Deep-Rooted Personality Traits

While some managers realize they must be strong minded, they are at the same time extremely sensitive to the dangers of a flawed decision. They are not sure of what is before them. Even if they habitually act with speed, the more they linger with the dangers of indecision, the more doubtful they become. "I habitually think of what I must do three or four months ahead; and I always look for the worst," declared Napoleon. Although Napoleon's actions were planned long in advance, he could also adapt successfully to the pressures of the moment and act swiftly and decisively. Making estimates beforehand, supported by all the available market intelligence, and thinking about your alternative courses of action are the best remedies for indecisiveness. You thereby prepare yourself to take clear-thinking action at any given moment.

Lack of Courage to Go on the Offensive, Triggered by a Manager's Innate Fear of Failure

Courage is the act of determination in a specific situation. It becomes a character trait only if it becomes a mental habit. Intellect in itself is not courage; there are ample numbers of brilliant managers who simply do not have what it takes to make timely and appropriate decisions. It is for you to arouse the inner feeling of courage and then act. You will have to face the critical moments when reason is pushed aside and replaced by the awful feelings that creep into your mind and take control of your actions. This is where training, discipline, and experience kick in to overcome those feelings. As an additional perspective, keep in mind that you are in a contest of mind against mind: your mind pitted against the mind of a competing manager who may be challenged by similar emotions. You want to be the one who prevails and moves forward.

Consider the extraordinary success of **Google Inc.** Much has been written about the bold and iconoclastic approaches that founders Larry Page and Sergey Brin exhibited when they started operations in 2000. Even then, they showed their bravado by going against the sage advice of seasoned consultants and analysts who advised selling out for a mere pittance during those early days. Industry watchers saw the behavioral patterns surface once again as the search giant quietly acquired a mobile-phone software company and began making forays into the instant messaging, Wi-Fi Internet, and telephony businesses.

Where, then, does courage come in? First, to maintain leadership in its existing business, Google has to keep innovating as the likes of **Microsoft** and **Yahoo!** actively search for competing market positions. Next, it takes serious doses of courage to enter new markets that are well protected by such giants as **eBay**, **Motorola**, **Nokia**, **SBC Communications**, and **Verizon**. Initially, Google had to play catch-up in those fields just to get a foothold. Then, the company had to find ways to carve out new or unserved market niches or to enhance its position with unique applications of technology.

No Trust by the Manager in Employees' Discipline, Capabilities, or Skills

This factor is most serious. The best laid plans, the most ambitious goals, and the most vibrant business strategies are not going to work with inexperienced employees who lack the essential business competencies. Also, if they do not display an implacable discipline or cannot demonstrate an aptitude for the job, the organization is in deep. As has been amply demonstrated in this and the previous rules, discipline, training, morale, and skill form the underpinnings of speed.

This ties to yet another contributing factor to support speed and the push for performance: the amount of time an individual stays in one position. As one executive at **Citigroup** explains, "There is kind of a natural evolutionary process where at some point people have been here a long time and they go off and do something else. That's healthy for an organization."

What is behind this statement? First, speed of reaction is needed at the lower echelons—from field personnel through midlevel managers—so that they can adapt quickly to the unexpected. Second, success in the marketplace is not a one-shot event. Rather, it is a serial process composed of many localized actions tied to objectives and market opportunities. Exploiting market situations depends on the intelligence and initiative of junior managers. They are the ones at the grassroots level most prepared to grasp the need for change, even where senior executives are reluctant to move away from their comfortable paths.

Third, long training and extended time at one job level may make managers experts in execution, but such expertise is bound to be gained at the expense of fertile ideas, originality, and flexibility—the essential elements for swiftly meeting the day-to-day demands of the marketplace. Junior managers, therefore, should demonstrate the qualities needed for speedy reaction. This is particularly relevant as the lean and mean organizational format takes hold and field personnel begin taking responsibility for on-the-spot decisions.

Consequently, if there is a lack of trust in some personnel, do not blame them entirely. Understand that a certain amount of attrition is desirable to maintain the agility, motivation, and energy to solve problems and identify moneymaking opportunities. It is also up to you to maintain an environment where knowledge, ongoing training, and discipline are bedrock components of successful performance.

No Trust by Employees in Their Manager's Ability to Make Correct Decisions

This is a particularly troublesome problem when employees' morale deteriorates and they lose confidence in their manager's ability to deal with what they perceive as a hopeless situation that could threaten their jobs—and even the company.

A semblance of that condition existed when Mark V. Hurd took over as CEO of **Hewlett-Packard (H-P)** in 2005. What caused H-P personnel to take a wait-and-see approach? Hurd had never managed a company as large as H-P. He walked into the mind-jarring job of restoring an $80 billion company to its former glory. Hurd had to determine if poor execution of a good strategy shaped by his predecessor, Carleton Fiorina, was the problem or whether the problem was good execution of a faulty strategy. Then, there was the big question about the flat growth of H-P's flagship printer business. Should that core product line be spun off, as some influential analysts suggested? Also, Hurd had to consider product and service innovations, the lifeblood of all high-tech businesses. What would it take to fire up H-P personnel to go back to the company's cultural roots to innovate and gain customer confidence? These are some of the high-minded issues that intelligent, experienced, and generally savvy H-P personnel wondered about before they would throw their hearts over the fence and embrace Hurd with trust.

Here, again, skill, courage, and determination are necessary traits that require continuous reinforcement. Where they are lacking, even excellent field managers often fail to perform if they reach higher executive positions. (Preliminary feedback indicated that Hurd scored with positive results.)

Inadequate Support from Senior Management

If gaps in communication exist among managerial levels, the result is inadequate command and control. In turn, such a vacuum prevents senior management from providing timely support in areas such as approving additional investment to increase market share, shifting resources to secure a competitive position, or improving a supply chain network. Also, guidance and support are needed from management on technology sharing, new product development, and personnel training. The all-important issue is to determine if marketing plans align with the company's strategic direction.

Disagreement and Open Confrontations among Line Managers about Objectives, Priorities, and Strategies

Where managers of equal rank cannot resolve difficulties independently, then it is up to senior management to intervene. In a multiproduct, multimarket organization, disagreement is typical and understandable, since most contentious issues deal with power and the availability of resources. This means that resources must be allocated among many business initiatives, all of which are vying for attention. Inevitably, some plans get short-changed or receive an outright turndown from management. Even the winners suffer the effects of lost time that slows their efforts. Painfully, the lapse in time may give the edge to an alert competitor.

Even at mighty **Microsoft**, there is intense competition for managers to tap its substantial resources. Beyond the issue of money, the company's three huge divisions often act as rivals—pursuing overlapping technologies and quarrelling over whose codes will prevail in the markets where their products interact.

A Highly Conservative and Plodding Corporate Culture Places a Drag on Speed

This is one of the toughest barriers to speed. Yet it is one that managers must face. If the culture is out of synch with the competitive environment and managers are not in position to change the culture, they must adapt plans to the existing culture with the aim of moving as rapidly as possible. At **Royal Dutch Shell** a sluggish culture existed due to the joint British and Dutch management structure, which plodded along with two chairmen and two executive committees. It finally took the courage of a new CEO to streamline the organization. One immediate priority was to speed up the "overly analytical culture" that made it difficult for the company to land big deals in a timely fashion.

Lack of Urgency in Developing New Products to Deal with Short Product Life Cycles

It is no secret that products exist in a short time frame and in a commercial world that clamors for faster, cheaper, and smaller. Where that basic knowledge is not fully internalized by management or acted upon quickly, results can be calamitous. The life cycle issue has gained so much importance that software developers such as **USG, SAP**, and **Dassault Systems** sell product life-cycle management (PLM) software. It is the fastest growing segment of the corporate software market, valued in 2005 at $5.9 billion and growing 14 percent faster than any other segment.

The PLM software allows people to coordinate their work within a specified stage of the product life cycle, whether they are in London, New York, or Shanghai. The system permits marketers to post ideas for new products. Engineers get to work designing three-dimensional prototypes. Manufacturers can lay out a new assembly line, complete with every piece of equipment necessary. With virtually no limitations, thousands of people can participate on a single project from anywhere there is a Web connection.

Organizational Layers, Long Chains of Command, and Cumbersome Committees Prolong Deliberation and Foster Procrastination

The essential ingredient for an efficient enterprise is simplifying the system of control and, in particular, shortening the organizational layers from the field to top-level executives. In a small organization, the chief executive officer or president is at the helm. He or she is in a unique position to control both policy making and execution. Because decisions do not have to be channeled through others, the CEO or president is unlikely to be misinterpreted, delayed, or contested. Plans can be implemented with consistency and speed.

In the larger multiproduct firms with more people, products, and additional levels of authority, results may fall victim to a cumbersome, inflexible operation. Individuals in the field often feel that there are obstructions in the decision-making process for moving into new markets. Missed opportunities are common, and "go" decisions get stuck for reasons other than the competition. Even first-line managers think that there are too many people at the staff level or not enough on the job with revenue divisions. The large office staffs and the shortage of line personnel are sources of constant complaint. Much of that condition is handled by downsizing and reducing staffs to an efficient "lean and mean" level.

Your own experience may well support the obvious inference that an organization with many levels in its decision-making process cannot operate with speed. This situation exists because each link in the managerial chain carries four drawbacks:

1. Loss of time in getting information back
2. Loss of time in sending orders forward
3. Lack of full knowledge of the situation by senior management
4. Reduction of the top executive's personal involvement in key issues that affect the availability of resources

Therefore, for greater efficiency and speed, reduce the chain of command. The fewer the intermediate levels are, the more dynamic the operation will be. The result is improved speed and increased flexibility.

A more flexible organization can achieve greater market penetration because it has the capacity to adjust to varying circumstances more rapidly. It can thereby concentrate at the decisive point before its competitors have a chance to respond. Forming cross-functional strategy teams made up of junior and middle managers representing different functional areas of the organization can further enhance organizational flexibility.

Aggressive Competitors Can Strike Fear among Employees, Damage Morale, and Result in Lost Momentum

Psychological impressions can create perceptions of success or failure in an individual's mind, as if they were actual physical encounters with competitors. "The psychological is to the physical as 3 to 1," declared Napoleon. Thus, perception—what the mind can conjure up, believe, and then react to—is indeed reality. Such perceptions feed employees' reactions, which generate a set of emotions. Some are correct; others are distorted. This means that employees will react to signs they choose to interpret and believe through observation, feel, or the inevitable rumors—right or wrong.

What follows is that employee morale could flounder based on the slightest signs of what employees perceive as a reversal. Such warnings could include failed performance of a new product, reduced profits, changing customer behavior, or competitors grabbing key customers. Then, there are the sudden changes in management structure, the resignation of a key executive, or the downsizing of operations that can send shivers through the employee ranks.

It is discipline, training, leadership, communications, and the underlying culture of the organization that will greatly influence how employees will react to these signs. Let us look more closely at each influence:

Discipline: Toughening-up employees to withstand the natural swings that exist in a global and competitive market takes a good deal of leadership skill and sensitivity to alter their attitudes and behaviors. This is particularly so when the natural inclination is to shrink from the realities of an uncertain and inconsistent environment. Therefore, your aim is to inspire individuals to make courageous save-my-company, save-my-job efforts and turn potential defeat into victory. Accordingly, maintain a continuing humanistic awareness of what the psychological effects can have on your business strategies and on your ability to manage your subordinates. For instance, if individuals are placed in a no-escape predicament and faced with the bleak outcome of dismissal, they are not likely to shrink from self-interest. Instead, they will often use their utmost creativity and energy to fight out of the tight spot. Thus, a real or perceived threat often creates remarkable behavior among individuals—if they are inspired by an energetic manager with a disciplined strategy for survival, followed by a plan for a burst of growth.

Training: There is the pragmatic truth that if employees are unaccustomed to the rigors of travel and long hours of work, they will worry and hesitate at the moment when levelheaded decisions are needed to handle tough competitive conditions. It is here that training and discipline support the payout of time and money.

Leadership: There are particular qualities that define a manager's ability to lead. These include insightfulness, straightforwardness, compassion, courage, and strictness, which are described in Rule 1.

Communications: What is needed is clear, uncluttered communications that convey information about strategic objectives of the organization, goals related to strengthening customer relationships, the long-term outlook about the market, and intelligence about competitors. Also, there are the tactical details that influence advertising, selling, and customer service objectives. Communications also refers to the internal structure of the organization whereby timely information is disseminated and decisions are implemented.

Culture: The company's culture appears as a set of behaviors and attitudes toward relationships with customers and suppliers. It is especially seen from the viewpoint of employees as intellectual assets to be nurtured and developed. Thus, indispensable to acting with speed and decisiveness is an organizational culture that helps all personnel acquire a way of thinking and an orientation that is totally customer driven.

Complacency or Arrogance as a Prevailing Cultural Mindset

An outlook that leans on a cocky attitude of "we've got the market locked up" often translates to complacency and ends with defeat. Such feelings tend to dry up creativity and stifle the inner drive to push for excellence. It siphons off the adrenalin that comes from fighting within a performance-based arena.

The marketplace is littered with one-time leaders in such fields as automobiles, electronics, metals, and consumer goods. Some companies in those industries turned around at the last moment; others were acquired or just went bankrupt. The mindset was usually supported by the inaccurate notion that high market share means security, which was often followed by a passive defense of that market position. Such conditions at one time or another faced such high-profile companies as **Xerox**, **General Motors**, **EMC**, and **Hewlett-Packard**. All, however, have aggressively pulled out of the slump or are in the process of attempting to do so.

Speed: A Core Rule of Strategy

Now, let us look at the big picture of speed as one of the core rules of competitive strategy. In the global market, the impact of speed is evident by how companies prioritize their long-term objectives and how they see their standing in the marketplace. In turn, those objectives filter down and impact operations in their local markets by regions, cities, towns, and villages.

Such is the case of Paju, a town in South Korea. Factories are being built at breakneck speed by the Koreans, Japanese, and Taiwanese in the fight to own the global television market. In particular, **LG.Philips** plans to spend a total of $25 billion in the next decade at Paju, where it will expand its plant eightfold. Meanwhile, **Samsung Electronics** will spend $20 billion to expand its liquid crystal display (LCD) panel operation. As fast, Japan's **Sharp Electronics** is erecting a $1.4 billion plant that will make large LCD glass panels in other parts of South Korea. This buildup in Korea and beyond has huge implications for the televisions consumers will be watching over the next decade.

With productivity and efficiency come lower prices, followed by the penetration of huge markets globally; this virtually assures those companies dominance in the television industry over several decades. Beyond television, LCDs are used for millions of computers, as well as an array of millions of cell phones, personal digital assistants, digital audio players, and cameras.

Summary

If you move with persistent speed, you can significantly improve your chances to secure a competitive lead. Apply speed in the following ways so that your timing preempts competitors' moves to frustrate your efforts:

- Explore opportunities to cut costs for you and your customers.
- Investigate strengthening quality assurance and introducing new warranties related to product performance and reliability.
- Speed up internal communications and decision making, particularly where it involves approval of products and services that are time sensitive.
- Track product life cycles for possibilities to replace products or systems and introduce new product applications.

In no case should you get bogged down in indecision or overlong, dragged out campaigns. Avoid procrastination. Instead, move forward and act with speed to maintain momentum and take on opportunities.

Before moving on, review Rule 3 using the strategy diagnostic tool to assess how this rule would affect your strategy.

* * *

Strategy Diagnostic Tool

Strategy Rule 3: Act with Speed

Part 1: Indications That Strategy Rule 3 Functions Effectively (Contributes to Implementing a Successful Competitive Strategy)

1. Managers and staff acknowledge that overlong, dragged out campaigns have rarely been successful. They also understand that half-hearted efforts divert interest, depress morale, and deplete resources.

    ☐ Frequently ☐ Occasionally ☐ Rarely

2. Managers realize that speed is an essential component to secure a competitive lead, with impact on market share, product position, and, ultimately, customer relationships.

    ☐ Frequently ☐ Occasionally ☐ Rarely

3. The staff understands that even minor delays can result in a loss of momentum and could signal a vigilant competitor to move in and fill the void.

    ☐ Frequently ☐ Occasionally ☐ Rarely

4. Managers and staff know that a strategy that integrates speed with technology is in a prime position to secure a competitive lead.
   ☐ Frequently ☐ Occasionally ☐ Rarely

5. Managers recognize that speed adds vitality to a company's operations and acts as a catalyst for growth.
   ☐ Frequently ☐ Occasionally ☐ Rarely

6. Managers understand that acting defensively to protect a market position is but a preliminary step to moving boldly against a competitor.
   ☐ Frequently ☐ Occasionally ☐ Rarely

Part 2: Symptoms That Strategy Rule 3 Is Functioning Ineffectively (Detrimental to Implementing a Successful Competitive Strategy)

1. Managers and staff fail to acknowledge that losses in market share and competitive position often result from prolonged delays when dealing with time-sensitive market conditions.
   ☐ Frequently ☐ Occasionally ☐ Rarely

2. A general malaise exists in the organization, which results in missed opportunities.
   ☐ Frequently ☐ Occasionally ☐ Rarely

3. Personnel lack initiative in implementing business plans.
   ☐ Frequently ☐ Occasionally ☐ Rarely

4. Managers are slow in preventing a product from reaching a commodity status.
   ☐ Frequently ☐ Occasionally ☐ Rarely

5. There is a lack of vitality due to staff inexperience and lack of training, which hinders growth.
   ☐ Frequently ☐ Occasionally ☐ Rarely

6. Managers are unable to act boldly and with speed, despite suitable market conditions.
   ☐ Frequently ☐ Occasionally ☐ Rarely

7. Managers are unable to secure a competitive lead due to sluggishness in integrating technology into the product and distribution mix.
   ☐ Frequently ☐ Occasionally ☐ Rarely

8. Organizational layers prolong deliberation, foster procrastination, and delay decisions.

   ☐ Frequently   ☐ Occasionally   ☐ Rarely

9. There is a lack of urgency in developing new products.

   ☐ Frequently   ☐ Occasionally   ☐ Rarely

The ratings for Parts 1 and 2 are qualitative assessments of managers' overall ability to use speed to execute an effective competitive strategy. Based on a diagnosis of your company's situation, use the following remedies to implement corrective action:

- Reduce the chain of command in your company or group (if possible) and increase the speed of communication from the field to the home office.
- Require junior managers to submit proposals with the prime objective of creating additional revenue streams through product innovations and new applications of technology, and by identifying new, unserved, or poorly served market segments.
- Use cross-functional strategy teams to benefit from multiple perspectives and gain buy-in to business plans.
- Gain input from managers and field personnel in locating competitors' weaknesses and areas of vulnerability.
- Conduct training to break down internal barriers to speed.

Remedies and actions: _____

_____

Finally, to demonstrate the far ranging impact of this rule, the following company problems linked to the use of speed come from a survey of chief executives of medium and large organizations (company names withheld to maintain confidentiality):

> "Products are late and customers are holding off purchases of existing products, which is slowing the company's revenue growth. The delays can only help the competition."
>
> "Delayed models damaged morale and created a dramatic loss of market share."
>
> "We must outmaneuver local and regional rivals by bringing sophistication more rapidly to the market than they can."

Strategy Rule 4

Grow by Concentration: Deploy, Target, Segment

Chapter Objectives

1. Use eight categories of markets to pinpoint segments with greater accuracy.
2. Shape concentration strategies for each market segment.
3. Implement tactical applications of concentration.
4. Use ethnographic guidelines to gain fresh insights into market behavior.
5. Employ the strategy diagnostic tool to assess Rule 4.

Introduction

Up to this point you have reviewed three strategy rules: shift to the offensive, maneuver by indirect strategy, and act with speed. What needs your attention now is an all-inclusive strategy rule that allows you to deploy your material, financial, and human resources to increase your chances of success. The rule is *concentration*. This classic principle is in sharp contrast to the overly common approach of spreading time and thinning out assets across numerous segments, objectives, and isolated actions—all of which would dramatically multiply your chances of failure.

What is behind this singular straightforward rule? Why should you concentrate all your resources at a customer group, geographic segment, or a single competitor? Are you not placing all your resource eggs in one precarious basket and taking on undue risk? Also, what impact would such a strategy have on how your firm is organized? For instance, do you centralize the sales force to handle all customers and products or do you divide them according to customer segments and product lines? What about senior executives and mid-level managers: Can they psychologically handle a concentrated approach or would they opt for the seemingly play-it-safe tactic of dispersing their resources over numerous segments, thereby increasing the possibility of being underweighted everywhere and strong nowhere?

The hardnosed evidence leads unequivocally to adopting a strategy that aims at concentrating resources where you can gain superiority in as few areas as possible. Using the broad guidelines of deploy, target, and segment, the following examples illustrate the application of concentration.

Case Examples

China is an outstanding instance that gives credence to the strategy rule: grow by concentration. With its enormous population of 1.3 billion speaking more than 100 dialects, China is about as diversified as any single country can be from a marketing point of view. It is in its diversity that concentration plays a central role in developing a viable business strategy. What people eat, wear, and drive differs greatly from north to south, east to west, rich to poor, young to old, city to countryside.

For the business manager and strategist, China is a superb example of how to use concentration to penetrate a market. Succeeding with concentration, however, incorporates a three-step process: (1) employ market intelligence to define customer segments and evaluate competitors, (2) use market data to pinpoint a segment for initial entry followed by a systematic rollout into additional market segments, and (3) customize products to those segments. Consider the following applications:

General Motors Corp. initially entered China by offering just a few car models, mostly large, high-end Buicks costing about $40,000. The primary market was fleet sales to government offices, which was singled out as a large and lucrative segment. After establishing a foothold and having developed a brand identity, GM managers began expanding into other market segments with selected models that matched customers' needs and their ability to buy. For China's fast growing rich consumers, there is the $75,000 Cadillac. Several notches down is the $30,000 Buick Regal for cost-conscious entrepreneurs who want a prestigious car. For midlevel managers, there is the $15,000–$20,000 Buick Excelle. For first-time buyers, there is the Chevrolet in several models ranging from $5,700 to $12,000. Specifically, for rural areas, a practical seven-passenger minivan, which can also haul an animal and a few sacks, goes for $4,000–$6,500.

Samsung discovered that its lines of appliances and electronics had to be directed segment by segment. Customers in the warmer climate of the south needed larger refrigerators than those in the more temperate north.

The domestic Chinese company **Haier Group** follows the same concentration strategy as it offers a range of washing-machine models, including a tiny one targeted at rural customers that costs just $37.

Procter & Gamble (P&G) has moved into China's rural areas with a budget detergent called Tide Clean White, while catering to its city customers with the more expensive Tide Triple Action. P&G has also developed bargain-priced versions of Crest toothpaste and Oil of Olay skin cream. This master of segmentation deliberately blends concentration coupled to marketplace needs to drive future product development, pricing, and packaging strategies. To that end, P&G relies on grassroots market intelligence from groups that live in the countryside. In addition, by conducting focus groups and surveys across the country, managers are immersing themselves in local cultures to determine how they can better tailor individual products at the initial development stage, rather than force-fitting existing brands to specific groups. Using ethnographic techniques (a tool discussed later in this chapter), P&G personnel go directly into the towns and villages, where they visit homes and farms and sit side by side with consumers, observe their lifestyles, and talk with them about product features they like and dislike.

Motorola Inc. discovered, after a few missteps, that it needed more information about its markets. Now it routinely sends teams of researchers into far-flung regions of China to determine needs and buying patterns. With the promise of a huge and developing segment, Motorola designers devoted more time and energy to the lower end of the market and found a ready market for its least expensive phones, which allow users to download MP3 songs and even customize ringtones. They also determined that consumers far from the large cities of Shanghai and Beijing are becoming more discriminating. Young people tend to look at value, although their purchases are very individualistic.

Rival **Nokia**, not to be left behind, produces scores of phone models aimed at every definable segment of user, with ground-level support provided through more than 100 sales offices in every corner of China. The strategic thinking behind the expansion of offices is the full realization that centrally managing markets will not work. Concentration by decentralization needs to be total.

The primary lesson that emerges from these cases is that to achieve an advantageous competitive position, concentrate in market segments that correlate with your core capabilities and link to your organization's strategic direction. Such segments would be found among overlooked customer groups with changing demographics and buying patterns or in segments that are beginning a fresh cycle of growth.

Consider the following example of an overlooked group: For a long period of time, marketers generally avoided the boomer generation—people born between 1946 and 1964. They assumed that the boomers' behavioral patterns and brand choices were hardened and that their biggest earning and spending years were behind them. Those assumptions proved wrong on all counts. Once awakened to the reality that a large and vacant segment was up for grabs, the more savvy managers began reexamining the boomers and came up with startling discoveries: Strong evidence revealed that this ignored group is far from sluggish, inactive, or set in its ways.

Instead, with average life expectancy at an all-time high in many developed countries, individuals over age 50 consider the middle age years as a new start in life and act as if they will be around for two or three decades. Surveys further indicate that few boomers plan to stop work altogether as they age. With children grown and out of the house, the lifestyles shape up like this: Second careers start, boomers strive to stay mentally and physically active, and they are open to new experiences and products.

All such intelligence signaled companies to move, and move rapidly, to gain a solid foothold in a previously disregarded segment that now exhibits outstanding promise for long-term growth. The promise is great, however, for those companies with the know-how and focused strategy to make opportunity a reality. Meaning: even with a potentially huge market destined to keep growing, the worst strategy is to attack the entire market. Instead, the concentration rule emphatically states:

> Avoid venturing into segments with products and services where you do not have unique and definable strengths. Otherwise, you squander budgeted resources and chase after niches that entail unnecessary risks.

Your best course of action is to focus on selected markets where you enjoy a measurable and substantial advantage over competitors; and where you have the know-how to sustain a lead through competitive strategies (see later section, "Eight Categories of Markets"). Then, match those parts of your business where you have an inherent and distinguishable expertise to the characteristics of the market segment you selected. Such a matchup becomes the underpinning for creating strategies built around product differentiation, value-added services, and business solutions that exceed those offered by your competitors (see Exhibit 4.1).

Look, too, within the supply chain for technology innovations that would create a market-driven advantage. For instance, check into next-generation computer-aided inventory controls, bar coding instrumentation, and just-in-time delivery systems. Included in the supply chain are the essential interactions with customers. In particular, watch how you perform on backup support in such areas as complaint

Exhibit 4.1 Concentration Strategy Applications

Controllable Areas	Strategy Options	Strategy Applications
Corporate culture	Understand your organization's culture through its dominant values, historical roots, and behavioral patterns. Also, understand that cultural values reflect in the practices and attitudes toward employees, as well as with the types of relationships formed with customers and suppliers.	Align strategies with your organization's culture. Also, use cultural guidelines to decode the actions of your competitors. The boldness or limitations of your strategy correspond directly with the prevailing corporate culture. (Refer to Rule 6 for techniques to strengthen or modify your organization's culture.)
Strategies and tactics	Successful strategies consist of five bedrock components: speed, concentration, indirect approach, alternative strategy, and unbalancing competition.	Introduce the five components of strategy into your business plans. (These essentials are discussed throughout this book.)
Personnel development	Developing personnel is accomplished through in-house seminars, public seminars, university programs, Internet degree and nondegree programs, one-on-one mentoring, self-study books, and interactive computer-aided instruction.	Personnel development is the human side of implementing business plans and coping with aggressive competition. Therefore, commit to ongoing training and skill development, with a focus on instilling employee discipline.
Product or service	Product options include quality, features, versatility, uniqueness, reliability, durability, style, color, design, brand name, packaging, and sizes. Options also include backup services, new technology, patent protection, warranties and guarantees, and compatibility with the Internet.	Use product options to concentrate on the specialized needs of customers, form a defensive barrier against competitive intrusion, and serve as a source for new products and services.

Exhibit 4.1 (continued)   Concentration Strategy Applications

Controllable Areas	Strategy Options	Strategy Applications
Price	Strategies include skim (high prices) and market penetration (low prices), as well as follow-the-market leader, market-driven, and cost-plus pricing. Strategies also take in pricing by segment and reacting to competitive actions.	Determine pricing strategies based on buying practices of the segment, competitive practices, and where products are positioned in their life cycles. Also determine market-share objectives, as well as where consumers are in their buying cycles. Applications are generally reflected by discounts, allowances, payment periods, credit terms, and financing.
Marketing/ promotion	Options include: - Advertising can be directed to the trade or directly to consumers. - Sales promotion can take place in the form of demonstrations, sampling, free trials, contests and sweepstakes, coupons, premiums, point-of-purchase and frequent-user inducements. - Sales force options cover basic compensation, incentives and bonuses, sales aids, sampling, amount of decision-making authority, level of training to maintain customer relationships, and ability to contain competitors. Internet uses include gathering information, selling products and services, and performing electronic transactions.	Apply marketing to: - Support sales force - Target prospects inaccessible to the sales force - Develop awareness and familiarity with brand, logo, package, or product - Improve supply chain relationships - Influence customer attitudes toward products and company - Carry out sales face to face or through direct marketing by mail order, telemarketing, call centers, and the unfolding applications of the Internet

Exhibit 4.1 (continued)   Concentration Strategy Applications

Controllable Areas	Strategy Options	Strategy Applications
Supply chain and physical distribution	Channel options cover the following pathways to the end-use consumer: - Wholesaler–jobber–retailer–consumer - Wholesaler–retailer–consumer - Retailer–consumer - Agent–retailer–consumer - Manufacturer–consumer	Strategy applications deal with the following issues: - Ability to maintain a viable position on the supply chain and ability to work effectively with intermediaries - Geographic size and market areas to be covered - Degree of channel control desired - Costs and ability to finance distribution - Capability to provide technical and service support and training - Type of delivery and inventory system required to achieve competitive advantage and customer satisfaction

handling and related customer services. The following case examples illustrate how companies are creating strategies to strengthen their market positions, increase market share, or prevent the loss of share.

Case Examples

Apple, **Hewlett-Packard (H-P)**, **Dell**, and **Gateway** are engaged in heavy competition concerning quality of customer support. At stake are the sought-after trio of customer satisfaction, loyalty, and brand image—all of which impact the bottom line, market share, and the future viability of the segment.

Dell, in particular, has worked hard to reverse a service slide in 2005. Service has been a major ingredient in the company's success formula, along with its unique strength of build-to-order convenience. That business model made the brand number one in computer sales in North America. Meanwhile, H-P and Apple launched several of their own initiatives by concentrating on designing unique diagnostic tools to help consumers detect their operating problems. Other organizations, such as **BMW** and **General Motors** looked at new-wave promotion trends through which they could carve out superiority. They began experimenting with lengthy ads of two minutes or longer. The ads incorporated more information and more drama, and used new technologies such as TiVo that let consumers watch TV or movies when desired. Other trend watchers, such as **Procter & Gamble**, **Miller Brewing**, and **Kellogg** have led the way by embracing the latest Web technologies, including social networking, podcasting, blogs, and interactive banner ads.

In addition to these examples, you will want to look inside your organization and concentrate on personnel skills. This is a definable area through which to develop a superior edge. The companies that are setting the gold standard for developing personnel and identifying talent include **IBM**, **General Electric**, and **P&G**.*
In particular, IBM's aim is to help managers analyze which skills their people possess and determine how those capabilities match the company's long-term strategic goals. IBM's approach: Train people ahead of anticipated changes.

Select Your Segment and Shape Your Competition

There are three tactical advantages to keep in mind when employing a concentration strategy. First, you create additional opportunities to target specific groups of customers by looking beyond the limiting approaches of classifying segments only by the traditional demographic and geographic criteria. (The balance of this chapter describes several unique approaches to segmenting markets.) Second, you are better able to maximize the impact of your marketing efforts against competitors. This means that you can avoid a head-on, direct, and costly confrontation through price wars and undifferentiated products, which will likely squander your resources and prevent you from reaching your goals. Third, using a concentration strategy helps you locate a point whereby you can pounce on your competitor's weakness and reduce its ability to respond to your actions. In turn, you are then able to gain the upper hand in one and possibly more segments.

---

* While these three companies represent the market leaders in their respective fields, it should alert managers in midsize and small organizations to take similar initiatives to the extent their resources permit.

Quick Tips: Rules of Concentration

1. You cannot be all things to all people. Even huge, global organizations will shy away from pursuing such a generally fruitless effort, unless they have sufficient resources to be a segment leader in all markets. (A mainstay of General Electric's strategy for decades has been to gain a number one or two position in a segment—or exit the market.)

2. Where there is an overwhelming leader in the market, your best strategy is to concentrate your strength and strike one point that represents your competitor's weakness.

3. Your best course of action is to acquire market intelligence that documents the specific needs and problems of a well-defined customer group and then concentrate your resources.

4. Look for a customer segment that represents a market trend or find a niche that has been overlooked, neglected, or poorly served.

To remain in a weakened position is not the hallmark of an effective manager. Your aim is to shift control in your favor. Therefore, as you prepare to deploy your sales, service, and other backup personnel, keep in mind that your intention is to wear down your competition and reduce the level of resistance facing you. For that reason, concentrating all efforts is the steadfast rule if you are to achieve an advantageous condition. Here, again, refer to Exhibit 4.1 to guide your thinking about how to determine areas by which you can assess your strengths, weaknesses, and, consequently, opportunities for shaping competition. The following case explains these points.

Case Example

Microsoft Corp. felt shock waves rippling through the organization during one recent period due to the aggressive moves of fast moving firms such as **Google, Salesforce.com, Linux,** and **Yahoo!**. These firms were actively involved in markets that Microsoft selected for future growth and diversification.

The giant company enjoyed decades of market dominance with its Windows and a suite of desktop applications for 80 percent of its sales. Now, with purchases in those markets reaching a plateau, Microsoft needed to expand into newer products, such as the Xbox videogame machine, MSN online service, wireless, and small-business software. However, Microsoft did not have enough time or deep enough market penetration to reach a commanding position. As of 2006, Microsoft was losing money in those markets. To some extent, it was a David versus Goliath encounter as the smaller and more nimble competitors put on a tough fight and were not intimidated by the giant intruder. It was Linux that took the lead in Web-server software in specialized segments. Google and Yahoo! commanded 70 percent search queries versus Microsoft's MSN with only 13 percent.

What options could Microsoft have used to develop concentration strategies and shape the competition? Here are some feasible choices:

- It could have flexed its superior financial and marketing muscles to overwhelm the smaller competitors. It could have focused its substantial technological resources on market segments with dedicated suites of problem-specific applications. It could have targeted lightly defended, poorly served, emerging, or overlooked customer groups.
- Other approaches might have taken the path of innovations in promotion, supply chain management, and pricing strategies.
- Organizationally, Microsoft could have developed greater flexibility by flattening the organizational hierarchy and creating market-driven business units that would react with greater mobility. The intent would have been to empower managers with increased decision-making authority and a grassroots capability to respond rapidly to market opportunities. (Microsoft subsequently made such organizational changes.)

Whichever approach Microsoft selected, a concentration strategy would likely have yielded positive results.

The next major area for shaping your competition is taking an intense look at competitors. Once again, go to Exhibit 4.1 to assess strengths and weaknesses, which would help you pinpoint areas for concentration. Also, get out and observe competitors' actual moves first hand. Use the ubiquitous Internet to obtain valuable information and, within budgetary constraints, use formal competitor intelligence methods. Collectively, you should be able to piece together a fairly accurate picture of your competitors' strengths and weaknesses. Complete the shaping process by selecting a target segment for entry. That translates to the familiar segmenting, targeting, and niching as the foundation approaches for implementing a concentration strategy.

However, as indicated earlier, the customary demographic and geographic approaches to segmenting markets are rather limiting, even superficial, if used exclusively for developing strategies. For best results look at segments under a high-powered lens that gives you greater insights into the character of the segment. While you can never foresee with certainty all its possibilities, you do get to recognize special qualities and get a better "feel" for the market. Consequently, the more you know about the proposed market segments, the more apt you are to devise effective strategies and tactics for initial market entry. As important, you ready yourself for an energetic follow-up should unexpected opportunities appear.

To assist in that effort, a relatively new application for business, *ethnography*, a tool widely used by anthropologists, is fast becoming an accepted approach to link products directly to consumers' wants and needs. The likes of **Nike**, **Apple**, and **Procter & Gamble** are actively involved in using ethnographic studies to design a product's features from its inception. **Intel**, for example, is staffed with a cultural anthropologist who uses ethnography and analytical techniques from other social sciences to "develop a deep understanding of how people live and work." Intel uses the knowledge acquired from these approaches to inform and guide the company's direction for technology applications and product development. Also, specialized software is beginning to appear that can add a great deal of sophistication to the effort.

As a starting point, you can immerse yourself in a grassroots approach to ethnography by using the following guidelines.

Guidelines to Utilizing Grassroots Ethnography

Step 1: Map a Segment

Define the physical dimension of an existing or potential market segment. Use personal observation, databases, and printed information to assess, verify, and define the segment. Record facts about the physical layout of factories, stores, warehouses, roads, or transportation hubs, as well as the locations of consumer groups within specific geographic areas. For an industrial sector, draw a map and annotate it with items that provide a clear picture of the area, such as period of architecture, condition of buildings, proximity of buildings to main roads, accessibility to railroad sidings, access to suppliers and services, condition of streets, and other physical details that are pertinent to your business.

For a residential neighborhood, draw a map of the area and comment on the physical condition and location of stores for purchasing daily essentials, availability of banks and similar services, types of residential housing, condition of schools, adequacy of street lighting, condition of main and secondary streets, and any other relevant physical details that would contribute to the validity of your segment selection.

Step 2: Create a Special Language

When sorting out the characteristics of a segment, avoid using descriptive terms that would tend to distort a picture of the behavioral makeup of the segment and thereby confine your thinking. One remedy is to create your personal descriptive language borrowed from a neutral source that does not influence you with pre-judged terminology, such as military, architecture, sports, or agricultural terms. The intent is to free your mind of biases.

Write down in narrative form your description of the segment. Include maps and other appropriate references related to demographics, geography, and behavior. This process will also help you redefine the makeup of an existing segment and even reveal fresh opportunities you may have missed by using conventional demographic studies. Your special language can release in you a fresh viewpoint to see groups, segments, and their respective cultures in a new light and as a new opportunity.

Step 3: Observe Body Language

Body language can communicate significant amounts of valuable information. The object is to equip you with a keen awareness of how to apply body language as a step in the ethnographic process for defining a segment with greater accuracy. Accordingly, learn to interpret meaningful gestures—for example, during prospecting and while observing the purchasing process, whether in a business-to-consumer or business-to-business environment. Write down your observations. The outcome should result in pinpointing buying patterns or in determining how deliberate or impulsive the buying decision is. In turn, such astute observations can influence how you promote to a customer, the amount and format of the information you put on a package, and the type and quantity of backup service you provide.

Also, where appropriate, observe usage patterns of the product or service (see previous case examples of companies selling into China). The central idea is to dispel stereotypical communications that result in inaccurate and costly assumptions about customer reactions and buying patterns.

Step 4: Describe the Ritual

This step is likened to the shopping and decision-making "ritual" practiced by a consumer, group, purchasing manager, or senior executive. As rituals vary with individuals, groups, and societies, so too do distinctive practices exist among consumers, companies, and various institutions.

Write an insightful description of the ritual. Describe in detail the physical setting and record (or flow chart) the events that make up the ritual, as in purchasing a product. There are no limitations about which rituals you observe. For example, they can include the purchase of sophisticated capital equipment, a computer system for a home, ordinary office supplies, life insurance, or home furniture. In all cases, you want to find out which practices are important to members of the group. This is a critical point if you are to grasp the viewpoint of the customer, which is the essence of relationship marketing. For that reason, stay flexible as you discover the key forms of behavior associated with an event.

Rituals include sentiments, emotions, and symbolic expressions that tie in to an event or occasion. Keep in mind, however, that rituals are built into the culture of a group or market segment. While it may be difficult to stay objective, try to avoid giving your personal opinion of some practice as right, wrong, strange, or familiar.

Finally, the purpose of this four-step process is to add greater precision to shaping your strategies. Therefore, make your observations and interpretations about the market segment you select meaningful. Indicate the type of product, service, packaging, pricing, and distribution you would recommend. The format in which this information is presented is usually a strategic business plan. As an additional aid in making the critical decision to enter a segment and concentrate substantial resources, refer to Exhibit 4.2.

Eight Categories of Markets

To augment your concentration strategy and provide a more comprehensive approach to market selection, classify markets as *natural, leading edge, key, linked, central, challenging, difficult,* and *encircled*. Using these eight categories as a guide, you can look with a more critical eye at what challenges you will face. Then, you will be better able to assess the risks and potential rewards when selecting your strategies. As you examine the characteristics for each market category, you may find some overlapping. That is acceptable since there are inherent commonalities among the various markets.

Natural Markets

In this type of market, a company operates in the familiar setting of its traditional markets. The inference is that, within such customary surroundings, personnel tend to be at ease and generally are not motivated to venture out of their comfort zone. To expand, they have to be motivated to move beyond the confines of existing markets—which gets you back to the point of organizational culture. That is, does your organization's culture permit venturing out of familiar territory?

For the most part, you and your rivals can operate harmoniously. That condition exists as long as each company sticks to its own dedicated segment. Generally, outright aggressive confrontations are seldom used. The primary reason for this uncharacteristic display of togetherness in a highly competitive world is that you and your rivals share a common interest in furthering the long-term growth and

Exhibit 4.2 Checklist to Determine Market Characteristics

Market Size
What size market can you handle with available resources?
Is there a segment in which you can concentrate with a distinct advantage?
In which segments do your competitors operate and what is the level of their penetration?
How is the market characterized: single market, regional market, multimarkets, national, or global?

Market Allegiance
How much commitment do competitors give to a specific segment: major, average, limited?
In terms of priorities, is there interest in the market's long-term development or in milking it for immediate gain?
How do these factors mesh with your firm's commitment?

Flexibility
Does the organization's culture permit a flexible response to unsettling market conditions?
How agile is your organization toward altering its strategies?
Is there a tendency to:
   Retreat from segments when demand weakens?
   Concentrate on key segments when demand increases?
   Harvest profits when sales flatten and let the market languish?
   Diversify into all new businesses or product lines?
   Aggressively challenge a competitor encroaching on your customers?

Positioning
How effective are you in carving out a defined position for a product or brand?
Where does your product fit on the product life cycle versus those of your competitors: introduction, growth, mature, or decline?
Is there room in the market for your branded product, as well as for a private label?
What is the likelihood of your product becoming a generic or commodity?

Product Usage
What is the rate and frequency of product usage?
How does the projected rate of usage rank with comparable offerings from competitors?
How adept are competitors in promoting more frequent product usage, finding new users, designing more product applications and solutions, or developing new uses for the product's basic materials or technology?
What is the frequency of introducing new or modified products?

Exhibit 4.2 (continued)   Checklist to Determine Market Characteristics

Pricing

What has been your strategy in pricing a new product?

Do you tend to use high (skim) pricing, low (penetration) pricing, or flexible (market) pricing?

What pricing strategies do competitors use?

Marketing

To what extent do you and competitors use advertising to support personal selling, inform target audience about the product, or bypass intermediaries and sell to end users through direct marketing?

How does the Internet fit into the total promotion mix?

How does the sales force compare with competitors regarding territory coverage, compensation systems, training, and technical and service backup?

How well do your competitors integrate sales promotion with advertising, the Internet, and the sales force?

Are there unique forms of communications in use that would affect your promotion, such as the prevalence of blogs and podcasts?

Supply Chain

How effective is the supply chain in reaching customer segments?

Are competitors displaying any strategies that could alter distribution methods?

Are any competitors attempting to control the supply chain?

What is the level of technology used in ordering, delivery, and inventory control systems?

Personnel

How do the levels of experience, expertise, and training of the competitor's personnel compare with those of your organization?

Is there a clear channel of communications from decision makers to those in the field?

How much grassroots responsibility and decision-making authority exist to handle sudden opportunities?

How would you describe employees' discipline when handling a customer crisis or dealing with aggressive competitors?

---

prosperity of the market. If any one company chooses to gain a meaningful benefit, however, a likely strategy might include securing a more advantageous position on the supply chain or adjusting its position by adding or deleting a link in the distribution network. Doing so would be sufficient to differentiate one company from a competitor and permit it to meet unfolding supply and demand situations with greater efficiency. Even with the mayhem of global rivals scrambling for market share or price wars eating away at profitability, it is still possible to have some semblance of market unity in natural markets.

There is one additional dimension that characterizes this category, which you should actively keep in the forefront of your thinking: Industries, markets, and products go through successive life cycle stages from introduction, growth, maturity, and decline to phase out. Much of the movement through these stages is driven by the adoption rate of technology, which precedes or follows changes in consumer behavior. There are also external changes triggered by legislative or environmental factors; these are generally out of a manager's ability to control.

Therefore, to maintain growth, take the lead in the search for new segments in which to concentrate. You can then provide the momentum that drives new product development. The following example illustrates these points.

Case Example

Hewlett-Packard enjoys a substantial presence in printers. For almost two decades, the $24-billion-a-year division has dominated rivals with a 50 percent market share. The central issue facing H-P management is that printer demand is flat.

Aroused by that dismal finding, managers woke with an urgency to find fresh ways to preserve profitability and propel the printer division forward. Beyond the usual cost-cutting approaches, managers developed an aggressive business strategy consisting of the following actions:

- Move out of underperforming niches.
- Cancel further projects that would entail initiating costly and time-consuming efforts slugging it out against competitors such as **Xerox**, **Canon**, and **Ricoh**.
- Concentrate a drive into offices by selling more affordable color laser printers.
- Roll out a radical new inkjet technology called "page-wide arrays" that uses thousands of nozzles to print an entire page at once.
- Push into fast growing consumer segments with a line of digital entertainment and photography products.
- Crack the huge commercial printing market with its unique high-speed Indigo printer by persuading the industry to replace older printing presses with faster digital devices that do massive runs.

Leading-Edge Markets

"Leading edge" means exploring markets by making minor penetrations into a competitor's territory. The intent is only to investigate the possibility of opening another revenue stream. Therefore, you want to acquire the following types of intelligence: (1) feasibility of the market to generate a revenue stream over the long term, (2) amount of investment needed to enter and gain a foothold in the market, (3) timeframe for payback and eventual profitability, and (4) assessment of competitors' market position and strengths and weaknesses.

As cited earlier, in what is now a classic example of a leading edge market, a few Japanese companies initially moved into the North American market with small copiers. **Xerox**, the market leader during the 1970s, concentrated its marketing efforts on large corporations with a line of large copiers. Xerox managers initially avoided the small copier market. That oversight proved to be a critical error because it allowed enterprising Japanese copier makers to exploit a wide-open opportunity to walk unopposed into the vast market of small and midsize firms. Once established, they moved upscale in a segment-by-segment assault and took over a significant amount of Xerox's primary market share.

Key Markets

"Key" means that you and many of your competitors seem evenly matched within key market segments. The general behavior is that you would not openly oppose an equally strong rival. However, should a competitor attempt to dislodge you from a long-held position with the clear aim of taking away customers or disrupting your supply chain relationships, then you may be forced to launch a countereffort by concentrating as many resources as possible on blunting the effort. Such actions are appropriate, however, if they fit your overall strategic objectives. Therefore, keep the strategic big picture in mind: If you expend excessive resources in hawkish-style actions such as price wars, then you may be left with a restricted budget to defend your market position. Therefore, shape your competitor, as discussed earlier.

Linked Markets

In this category, you and your competitors are linked with easy access to markets. Your best strategy is to pay strict attention to constructing barriers around those niches that you value most and from which you can best defend your position. Barriers include:

- Above-average quality
- Feature-loaded products
- First-class customer service
- Superior technical support
- Competitive pricing
- On-time delivery
- Generous warranties
- Patent protection

Not only do these build barriers against competitors' incursions, but they also go a long way in solidifying customer loyalty. In particular, customer loyalty gives you a long-lasting, profit-generating advantage that is difficult for a competitor to overcome. It is the one area that makes a meaningful addition to your growth. As one management analyst put it: "If you currently retain 70 percent of your customers and you start a program to improve that to 80 percent, you'll add an additional 10 percent to your growth rate." The following example illustrates how one company, with about equal access to the market with rivals, dealt with customer loyalty.

Case Example

Samsung Electronics Co. manufactures high-definition TVs, phones, plasma displays, and digital music and video players. Its market concentration is in low-margin consumer electronics. Further, Samsung favors hardware over software, which defies generally accepted industry doctrine. That one aspect of its strategy creates a defensive barrier and fortifies its special area of concentration, which is remote from other competitors.

Samsung surged ahead, making remarkable progress over a six-year period from a once financially drained company saddled to a brand name associated with cut-rate, me-too TVs and microwaves. It now wins design awards and produces feature-loaded products, which is rapidly moving the company to the top of consumer-brand awareness in big-screen TVs, cell phones, flash memories, LCD displays, DVD players, and dynamic random access memory (DRAM) chips.

From virtually nowhere, Samsung has risen as a market leader across the technology range. It is doing so by maintaining a strategy that defiantly refuses to enter the software business of music, movies, and games, as **Sony** and **Apple** have done. Instead, concentration remains strong as the company's vertically integrated manufacturing strategy. It also distinguishes itself by avoiding the typical industry approach of outsourcing production.

All told, Samsung's strategies cover the major rules already discussed in this book—namely, shift to the *offensive*, act with *speed*, and grow by *concentration*. Specifically, Samsung:

- Cut into bureaucracy so that fewer layers existed as barriers to timely approval for new products, marketing budgets, and business plans. The reorganization also resulted in speeding the decision-making process and increasing managers' ability to seize market opportunities and latch onto emerging buying trends.*
- Created a highly competitive internal working environment and concentrated efforts on business solutions anchored to design, function, features, and cost.

---

* Organizational design and the value of shortening the lines of communication are detailed in Rule 3: Act with Speed.

- Customized as much as possible. For instance, in memory chips, instead of getting caught up in a commodity price battle, Samsung commands premiums that are 17 percent above the 2005 industry price average. The primary reason is that 60 percent of its chips are customized for such products as **Dell** servers, **Microsoft** Xbox game consoles, and **Nokia** cell phones.
- Maintained speed. What used to take 14 months to go from new product concept to rollout now averages only 5 months. To achieve that level of efficiency, Samsung maintains an extremely flexible organization.
- Stayed on the offensive. The bold strategy of rapidly deploying personnel also permits the company to refresh its product lineup every nine months, which beats competitors' capabilities and results in a credible market advantage.

Central Markets

"Central" means that a company faces powerful forces that threaten its superior market position. These forces are as diverse as watching small companies niche away at a market leader's position through aggressive pricing or offering dazzling feature-laden products, or by looking at technology-rich firms generating new applications overlaid with value-added services. To counter such threats, firms use joint ventures so that the cumulative effects yield greater market advantages and offer more strategy options than can be achieved independently. For many companies the merger and acquisition (M&A) route and other forms of joint ventures have proven the strategy of choice. The following industry example illustrates this category of market.

Case Example

The banking industry has been consolidating for well over a decade into the hands of several mega banks, dominated by the likes of **Citigroup**, **J. P. Morgan Chase**, **Bank of America**, and **Wells Fargo**. During the high-profile maneuvering of the giants, small banks fell by the wayside and remained in a somewhat dormant state.

Then a fissure slowly appeared among the big banks into which a few surviving smaller banks forced a cavernous opening. The crack first became visible to executives at smaller banks as growing numbers of retail customers and commercial borrowers complained about long waiting periods for loan approvals, increasing account minimums, rising ATM fees, and declining customer services. Taking advantage of the opening, small banks charged forward offering scorned customers the royal treatment with an extensive variety of friendly efforts, from serving Starbucks coffee, free babysitting, investment advice, no minimum balances, and customized account services to very fast approval of loan applications.

In effect, the small banks played their winning hand by exposing the megabanks' inability to respond appropriately with superior service, reach out to underserved niches within minority communities, open branches in areas abandoned by big banks, or cultivate loan terms backed by the personalized attention from senior-level bank executives. The result is that deposits at small banks have grown by 5 percent a year since the mid-1990s, while growth at large banks has been flat; profits have grown at 11.8 percent annually versus 8.5 percent for the big players.

The major banks did not generally respond suitably during that period. They continued to pursue very focused strategies, dominated by maximizing profits through cutting costs and similar financially oriented measures.

Challenging Markets

In this category, if you enter a market dominated by a strong and aggressive competitor, be watchful. You could place your company at excessively high risk. If, however, your long-term objectives strongly support maintaining a presence in a challenging market and the expenditures of financial, material, and human resources are consistent with your overall strategy, then find a secure position on the supply chain. It could be one of your single best chances for lessening the risk and achieving a solid measure of success. Your aim is to rely on efficient distribution to ensure the movement of finished products to customers.

Dell Computer is a prime example of excellent supply chain management. The company activates its manufacturing process and the supply chain only when an order is received from a customer. That strategy eliminates the cost of storing excessive inventory. Dell benefits by shipping just the right amount of components to its factories, thereby avoiding investing in expensive warehousing. For instance, in one facility, what used to be done in more than two buildings now is accomplished in one by applying the techniques of supply chain management.

Difficult Markets

This type of market is characterized as one where progress is erratic and highly competitive. If you attempt to make any meaningful market penetration, secure key accounts, or maintain reasonable levels of logistical support, you are likely to be blocked by asset-draining barriers. Also, if a competitor is fully prepared, takes you off-guard, and causes you subsequently to lose your market position, it is difficult to return to your former position. In effect, you are entrapped in an untenable condition and your entire business strategy could be in jeopardy. Your best course of action is to go forward, as long as the effort is consistent with your mission and long-term strategic objectives. Here is how one company maneuvers through its difficult market.

Case Example

Liz Claiborne Inc. is an apparel company skillfully balancing 26 brands that attract consumers spread over the demographic and psychographic (behavioral) spectrum, from teens to middle-aged women to bargain shoppers. The company is the master of niche marketing and branding in a highly competitive and difficult market. Obsessively anchored to research trends and sales data, Claiborne's business practices have ingeniously assumed the status of a science in a business traditionally bent on fashion by inspiration, whim, and attempting to make trends rather than following them.

Claiborne relies heavily on initiating numerous consumer studies, hiring color- and trend-consulting firms, and even utilizing a small research firm staffed by psychologists who study women's shopping behavior, going so far as to comb through their closets. Armed with reams of data, Claiborne's 250 designers methodically interpret the market trends for their respective clothing labels, which include DKNY, Lucky Brand Dungarees, Shelli Segal, Kenneth Cole, Dana Buchman, Villager, and Crazy Horse. Nine additional brands play off the Liz Claiborne name. Thus, the customer profile and buying behavior associated with each brand are carefully dedicated to the market niche in which each designer operates.

Does the system stifle creativity? Apparently not. In a chaotic market where industry sales plummeted 7 percent during one period, revenues at Claiborne jumped 11 percent. During the following sales period, revenues skyrocketed 66 percent. Dedicated market research continues to drive the business and permits Claiborne to prosper in a market that is as accessible to the giants as to numerous boutique firms.

Encircled Markets

An encircled market foretells a potentially risky situation. This market condition exists when you control limited resources and any aggressive action by a stronger, well-positioned competitor can force you to consider pulling out of a market. Therefore, it is in your best interest to maintain ongoing competitive intelligence to assess the vulnerability of your position accurately against that of your opponent. Armed with the intelligence, you can then develop a contingency plan that highlights your strengths and exposes your competitor's weaknesses. If, in your judgment, you still lack maneuverability and a capability to mount a meaningful competitive response, then exiting the market is a prudent way out, as long as it minimizes disruption to your main line of business.

If, on the other hand, your competitor foresees an untenable position, it is wise to give the rival a way out of the market and not force him into a fight-to-the-end mindset. Strategies include: (1) exploit a competitor's weaknesses and aggressively stay ahead by developing product enhancements, (2) launch value-added services, and (3) initiate any other programs that would hamper his ability to maintain a profitable market position. The aim of these strategies is to discourage your opponent from making a monumental effort to survive. Instead, encourage him to take the more tempting approach and exit the market.

To implement the three-part strategy, it is best to form cross-functional teams. The following examples illustrate the advantages of this organizational design.

Case Examples

Motorola uses teams consisting of engineers, designers, and marketers who work in an innovation lab known as Moto City. One noteworthy result: the RAZR, a half-inch thick, ultralight cell phone that sold a breathtaking 12.5 million units within less than a year of its introduction.

The outstanding feature of the team approach is that it creates a seamless connection from developing a product concept to satisfying the needs of the market. Additionally, it cuts development time and eliminates a fair measure of risk by delivering products to the market in a timely manner. Further, with product life cycles getting compressed into shorter time periods, the team is able to tune in to the marketplace with remarkable agility.

Mattel's preschool toy unit, **Fisher-Price**, uses a similar team approach to cut through the multiple tier corporate bureaucracy. Called the Cave, the exceedingly casual center located far from headquarters boasts teams of staffers from engineering, marketing, and design who meet with child psychologists or other specialists to share ideas.

After observing families at play in the field, they return to brainstorm, from which prototypes of products are built. In all, mingling with people from various disciplines has been a vital key to the teams' product successes.

To recap, familiarizing yourself with the eight categories of markets (Exhibit 4.3) will equip you with additional insights when selecting markets to enter and strategies to pursue. Further, if you are able to concentrate your resources, rather than create an unequal distribution that dissipates your strength, you are in an excellent position to exploit your opponent's weaknesses successfully. By that means, you conserve resources for unexpected opportunities that would otherwise be denied to you. It also permits you to defend against an unforeseen attack from a competitor. Such thinking and positioning are marks of an effective manager.

In order to make the most of concentration in your business strategy, use the following guidelines:

1. Use competitive analysis to identify your competitors' weaknesses or market gaps. Doing so gives you the opportunity to shape your competitors.
2. Concentrate on an unserved, poorly served, or emerging market segment that represents growth and, in turn, could help you launch into additional market segments.
3. Introduce a product (or product modification) not already developed by existing competitors that can fortify your position and reinforce your concentration strategy.
4. Secure your primary segment by private labeling your product concurrent with establishing your own brand. You thereby create a barrier by blocking a competitor's entry.
5. Expand into additional market segments with new product offerings. Also consider modifying existing products, providing value-added services, building in specialized applications, or developing creative packaging.
6. Recognize that every market segment you enter is actually the starting point of a new operation and a new sales cycle, which opens up a whole set of fresh possibilities.
7. No campaign is complete without first determining where to concentrate your efforts. The aim is to apply maximum effort at a decisive point. That point means positioning your resources to satisfy market demands, while preempting your competitor from taking similar action with any chance of success.

Finally, when you are developing a concentration strategy, you should address certain questions:

1. *Should you change the allocation of your resources after you have gained a favorable market position?* If you have gained a dominant position, as in the **Apple** iPod case with 85 percent market share, then you should move partially from the offensive to the defensive, making certain that you have an active defense.* You saw the effect of a somewhat lethargic response by Xerox's management when Japanese copier companies first showed interest in the North American market and then moved energetically to concentrate in the small-business segment as its initial point of entry. Consequently, how you deploy your resources and, in particular, how you shape your marketing mix should change with the situation. This also means adding as much flexibility to your organization as feasible, which includes holding reserves. You thereby ready yourself to respond rapidly to counter competitors' moves.

    Also, one of the essential activities related to allocating resources properly is to acquire as much competitor intelligence as possible. It is from such information that you pinpoint funds for products that could successfully

---

* As of this writing, IBM and Sony have announced competing products. Others are sure to be on the way.

Exhibit 4.3 **Eight Categories of Markets for Concentration**

Category	Characteristics
Natural market	You and your rivals can operate harmoniously as long as each company sticks to its own dedicated segment. Generally, outright aggressive confrontations are seldom used. All companies share a common interest in furthering the long-term growth and prosperity of the market.
Leading edge market	Market entry means a minor penetration into a competitor's territory to determine the feasibility for generating a long-term revenue stream.
Key market	Competitors appear evenly matched within key market segments. The general behavior is that you would not openly oppose an equally strong rival. If the competitor attacks your position, then you are forced to launch a countereffort by concentrating as many resources as necessary to blunt the competitor's attack.
Linked market	You and your competitors are linked with easy access to markets. Your best strategy is to construct barriers around those niches that you value most and from which you can best defend your position.
Central market	You face powerful competitors that threaten your market position. Counter such threats by joint venturing with firms so that you gain greater market advantages and develop more strategy options than you would be able to accomplish independently.
Challenging market	An aggressive competitor dominates the market and thereby could place your company at excessively high risk. However, if your long-term objectives strongly support maintaining a presence in the market, then one prudent approach is to secure a solid position on the supply chain.
Difficult market	Competition is pervasive and market behavior is erratic. Gaining and maintaining market penetration are difficult. Overall, your best course of action is to go forward only if the effort is consistent with your mission.
Encircled market	This market is risky; any aggressive action by a stronger, well-positioned competitor can force you out of the market. Maintain ongoing competitive intelligence to assess the vulnerability of your position against that of your opponent accurately. If you lack a capability to mount a meaningful competitive response, then exiting the market is a prudent strategy.

challenge competitors' entries. You are also able to decide how to deploy sales people, motivate middlemen, and budget promotional money. The bottom line is that one of the advantageous outcomes is that as you concentrate your resources, you thereby put a competitor in a position of spreading his efforts and weakening his overall market position.

2. *How will your personnel react under diverse market conditions and during actual competitive confrontations?* The straightforward answer is that only the skilled will survive. The actions of personnel under a variety of circumstances should continuously remain top of mind and your primary concern. Therefore, the quality of your individuals is far superior to quantity and you should not sacrifice quality. If you do, there is failure unless the competition is far inferior to yours. While no situation offers certain results, it is axiomatic that the skilled and watchful eyes of highly trained individuals can turn disadvantage to advantage and prevent an advantage from turning into a setback.

   Conversely, the unskilled will likely fail. Where, then, does the ultimate responsibility lie to assure a prompt and successful response to a competitive challenge? Plainly, accountability for bottom-line results remains with *you,* the manager. Training your personnel to act promptly and correctly so that they do not falter is an essential factor in leadership and managerial competence.

3. *Are you prepared for changes in your personnel's behavior if the competitive situation shifts from success to failure?* Determine from past performance how your personnel behaved before, during, and after a tough market encounter. For instance, were there signs of order or disorder if business plans started to crumble? Specifically, was there a point at which personnel "lost it" and were ready to pull away? Was there an instance when managers were able to resume control? Also, how were managers perceived by those they supervised? Find out if communication channels were intact and whether they led to understanding or confusion. Find out the overt or disguised messages that generated encouragement or discouragement, elation or fear. What can be said about all these points with reference to the competitor?

4. *Since the best long-term strategy cannot produce good results if short-term tactics are at fault, how do you assess your overall performance compared with that of your competitor?* Look at how first-line and midlevel managers performed under a variety of market scenarios. Also determine how the next higher level managed them. All too often, it is not the underlings that are at fault; rather, upper-level executives must take full blame in such areas as incompetent leadership, poor or nonexistent market and competitive intelligence, inadequate or unclear communications, ignorance of fundamental strategy rules, and an inability to motivate employees to implement plans with speed and enthusiasm.

The following case summarizes the strategy rule to grow by concentration.

Case Example

Marriott International Inc. has done an excellent job of targeting its several hotel brands—Courtyard, Renaissance, and Marriott—to precisely defined groups. During its primary growth years, the mainstay of Marriott's brand image was built on outstanding service. Most recently, however, challenged by changing behavioral trends and pushed by a strong and highly visible customer segment of younger business-casual and iPod-playing travelers, Marriott decided to go for a makeover and target that customer group.

In that niche, Marriott attempted to play catch-up against a smaller and more nimble rival, Starwood Hotels, which has moved toward postmodern room designs. Marriott introduced newly remodeled hotel rooms for its flagship hotels using a backdrop of music and comedy that captures an entertainment theme. Further, Marriott followed through with ads aimed at the road warriors who are attracted by technology-friendly surroundings. To implement the changes, Marriott recruited young brand managers from **Nike**, **Procter & Gamble**, and **Coca-Cola**.

Marriott owns a mere seven hotels and manages or franchises the rest of the vast chain, so management's toughest job was persuading franchisees to pay for the upgrades. While some complained, the viable choices were few: change or be left behind.

The essential points of the case are:

- Determine your business unit's or organization's strategic direction as it relates to evolving market trends and changing customer behaviors.
- Select segments that represent the best possibilities for growth.
- Shape your strategy so that it satisfies the needs of your customer segments, yet avoids a direct confrontation with competitors.
- Concentrate your efforts according to your material, financial, and personnel capabilities.

Summary

Follow these guidelines to grow by concentration: Stay with a strategy of concentration. It is your best approach rather than spreading your resources too thin, which will only add more areas of weakness and expose additional points of vulnerability. Also, maintain scrutiny over your markets through competitive intelligence (see Rule 5 in the next chapter). It is inconceivable that hungry competitors will sit still and leave sales and profits on the table for only a single company to devour. In particular, that truism also applies to flat or slow-growth markets, where any sales increases will usually come from unsuspecting competitors. Related to this, trying to regain market share is always costlier in expenditures of time and resources than fighting hard to hang on to the market share you already possess. Also, valuable resources have been spent to obtain share in the first place. Therefore, why pay twice to regain the same position?

Look, too, at developing a flexible organizational design that allows for a quick response to a competitive threat—that is, make effective use of rapid communications through the variety of new technological gadgetry. To the extent that your authority carries, try to reduce the layers of management that hamper quick decisions. In their place, introduce cross-functional strategy teams with decision-making authority (see **Motorola** and **Mattel** examples).

If you face an alarming competitive situation as described in the **Xerox** case, it is in your best interest to develop contingency plans laced with what-if strategies to cover unexpected challenges by aggressive competitors. Also, conduct training programs at all managerial levels that address a range of real-world competitive threats. It is difficult to expect your personnel to behave with any bravado through the pounding pressures triggered by aggressive competitive actions. The strain of suddenly going from an offensive to a defensive mode creates a jarring experience. It can be so severe that untrained individuals will succumb to fearful emotions whereas a bold and disciplined response is needed.

Before moving on, review Rule 4 using the strategy diagnostic tool to assess how this rule would affect your strategy.

* * *

Strategy Diagnostic Tool

Strategy Rule 4: Grow by Concentration

Part 1: Indications That Strategy Rule 4 Functions Effectively (Contributes to Implementing a Successful Competitive Strategy)

1. Managers are competent at concentrating their resources to gain a superior position in a selected segment.

    ☐ Frequently   ☐ Occasionally   ☐ Rarely

2. Managers use market intelligence to pinpoint a segment for initial market entry. Then they use that position to roll-out toward additional growth segments.

    ☐ Frequently   ☐ Occasionally   ☐ Rarely

3. Managers are adept at going beyond the traditional demographic and geographic segmentation approaches by employing more finite classifications (as suggested in this chapter) to identify new or underserved segments.

☐ Frequently ☐ Occasionally ☐ Rarely

4. Managers are flexible about moving out of underperforming segments and concentrating on faster growing segments.

☐ Frequently ☐ Occasionally ☐ Rarely

5. Managers are proficient at concentrating their resources against competitors' weaknesses.

☐ Frequently ☐ Occasionally ☐ Rarely

6. Marketing and sales personnel are skilled at focusing products and services to suit the specific needs of their customers.

☐ Frequently ☐ Occasionally ☐ Rarely

7. Managers recognize that concentration is a workable strategy to challenge larger competitors on a segment-by-segment approach.

☐ Frequently ☐ Occasionally ☐ Rarely

8. Managers understand that every market segment presents opportunities to fill market gaps, allocate resources efficiently, and exploit a rival's limitations.

☐ Frequently ☐ Occasionally ☐ Rarely

Part 2: Symptoms That Strategy Rule 4 Is Functioning Ineffectively (Detrimental to Implementing a Successful Competitive Strategy)

1. Managers are not able to accurately select market segments that represent long-term growth.

☐ Frequently ☐ Occasionally ☐ Rarely

2. Managers tend to dissipate resources across numerous segments.

☐ Frequently ☐ Occasionally ☐ Rarely

3. The company experiences numerous failed product launches by not explicitly building its new-product strategy around segment needs.

☐ Frequently ☐ Occasionally ☐ Rarely

4. Managers have not internalized the rule that to achieve an advantageous competitive position requires a strategy of concentration, rather than spreading resources and exposing pockets of vulnerability.

☐ Frequently ☐ Occasionally ☐ Rarely

The ratings for Parts 1 and 2 are qualitative assessments of managers' overall ability to use concentration to execute an effective competitive strategy. Based on a diagnosis of your company's situation, use the following remedies to implement corrective action:

- Install an ongoing competitor intelligence system to identify a competitor's weaknesses.
- Concentrate on an emerging or poorly served market from which to expand into additional segments.
- Secure your position with dedicated services and customized products that would create barriers to competitors' entry.
- Within market segments, tailor products and services built around product differentiation, value-added services, and business solutions that exceed those of competitors.
- Conduct strategy training sessions, especially for those individuals who resist adopting a strategy of concentration.

Remedies and actions: _____

_____

Finally, to demonstrate the far ranging impact of this rule, the following company problems linked to the use of concentration come from a survey of chief executives of medium and large organizations (company names withheld to maintain confidentiality):

"Need to develop an organization to yield a new and far more competitive company."

"We need to develop small, quasi-independent units, each of which focuses on its particular market with freedom to do what it takes to win."

"We must drive down costs, while beefing up R&D and finding new places to grow."

"We have to find ways to woo back angry retailers and build the brand image among groups that were once behind the product."

Strategy Rule 5

Prioritize Competitive Intelligence: The Underpinnings of Business Strategy

Chapter Objectives

1. Utilize the tools and techniques of competitive intelligence.
2. Identify the behavioral "personalities" of competitors.
3. Select agents to augment traditional competitive intelligence techniques.
4. Apply competitive intelligence to develop your business strategies.
5. Employ the strategy diagnostic tool to assess Rule 5.

Introduction

Think about one of your more serious clashes with a competitor. At what point before, during, or after the encounter could you answer any of the following questions with a fair degree of accuracy?

- *About the competitive problem*
  - When and how did you receive the first signal about your competitor's aggressive moves?
  - At what point did you fully grasp the seriousness of what the competitor could possibly do to your market position?
  - Who noticed the first signs of a problem: an experienced executive, a sales rep, or a novice manager? Looking back, how accurate was the assessment?
  - How confident were you of its reliability? Did you consider the assessment useful enough to commit company resources?
- *About your competitor's performance*
  - If your competitor's strategy failed, at what point in the unfolding action did you begin to see his efforts come apart?
  - Did the competitor leave the market, mount a completely new effort, or retreat to his starting position?
  - What were the telltale signs of weakness that caused the competitor's poor performance?
  - Did the strategy fail because of insufficient financial or physical resources, lack of trained and experienced personnel, weak leadership, or faulty implementation?
- *About your company's performance*
  - If the competitor was successful, at what point did your own situation unravel?
  - Did the competitor catch you by surprise?
  - What, specifically, was the nature of the competitor's strategy that caused your failed performance?
  - Was it a case of insufficient reserves, manpower, budget, or backup support or did you wait too long to react?
  - On the other hand, if your efforts were successful, did you do an after-action report to determine how you succeeded, in what timeframe, under what conditions?
  - When, exactly, was the turning point that assured your success?
  - Was there an instant where events could have gone awry because of insufficient intelligence?
  - How did your personnel hold up during the action?
  - What lessons did you learn about your staff, your strategies, and the reliability of your competitive intelligence from the unfolding events?
  - Would they be topics for formal training or informal discussions at meetings?

The point of these questions is that when you are immersed in a tough competitive encounter, you may not be fully acquainted with the overall strategies and short-term actions of your competitor. For that matter, you may not be totally clear about your own strengths and vulnerabilities. Such doubts could include the capabilities of your personnel to perform under adverse conditions.

This is not surprising. Trying to predict the performance of individuals at a particular time and during a special experience is fraught with uncertainties. Individuals blinded with fear and apprehension hinder calm and factual assessments and often fill reports with inaccuracies. Also, people tend to believe evil rather than good. Consequently, what you get is feedback tainted with bad news and false dangers. Even with what you believe is reliable intelligence, the moment you begin carrying out your plans, a thousand doubts may arise about the dangers and pitfalls that could develop.

Invariably, the situation affects you, too. Gnawing misgivings creep into your mind about being seriously mistaken in your estimates of the situation. Such feelings of uneasiness often take hold and from uneasiness to indecision are small, scarcely discernible steps.

What are the underlying issues that trigger such active thinking and subconscious feelings? Are they doubts about the competitor's strength, exaggerated estimates of its size, and concerns about the accuracy of the information? Specifically, what about those individuals gathering and reporting the intelligence? As indicated before, individuals by nature are inclined to be timid and tend to overstate danger or, just as likely, they steer toward self-interests as their prime motivation. All these issues can weigh heavily on your mind and leave a misleading impression of your rival's intentions and capabilities. From this mindset arises a new source of indecision.

In spite of these complexities, you cannot take uncertainty as a reason for indecision or as a rationale to forego action. Your best bet is to prepare as best you can by elevating competitive intelligence to the highest, *highest* level and making it an embedded part of your strategy development and its implementation.

Competitive Intelligence: The Underpinnings of Business Strategy

The central idea behind competitive intelligence (CI) is this: If you know your rival's plans and are able to monitor his moves, then you can estimate with some degree of accuracy which strategies are likely to succeed and reject those with minimal chances of success. You can also assess the level of risk associated with each of your options. Further, with CI you can define your competitor's operating patterns. You can determine his market position, deployment of personnel, and where you are likely to face the most or least opposition. Knowing where your competitor's strength is formidable and where it is weak gives you lead time to take counteractions. What better way is there to establish personal credibility for your managerial skills than being able to motivate your staff to handle potential threats and opportunities?

Therefore, CI as an information-gathering, decision-making tool affects all operating parts of a business either directly or indirectly. It is the centerpiece of all offensive and defensive actions. By exposing strengths and weaknesses in your situation, as well as in those of your competitors, CI functions as the core component when you are developing strategies and tactics. If utilized with the same care you would give to a delicate instrument, CI can signal subtle changes in the marketplace. For instance, it can help you preempt and outmaneuver competitors, preserve financial expenditures for the most expeditious timing, and even protect your corporate secrets from inquisitive onlookers.

Therefore, listen to the incoming intelligence and be sure to evaluate the reliability of each source. Then, you can validate or disprove reports. In either case you gain confidence and are able to make appropriate decisions.

The Call to Action

After you have assembled reliable intelligence and carefully deliberated on all that is meaningful, develop and implement your strategy with speed. It would be a far greater error to wait for a situation to clear up entirely. The reality of working in a dynamic competitive marketplace is that decisions are required, even in the fog of uncertainty.

Therefore, feel assured that the competitive intelligence you have assembled and screened and the assessments you have made will produce the results you expect. Even in the penetrating light of reality, if you discover that some intelligence is contradictory, false, or doubtful, then you have little recourse but to lean heavily on your judgment and move on—that is, depend on your knowledge of the industry, trust in your years of experience, show confidence in those key individuals with whom you work, recognize the value of your formal and informal training, and rely on the richness of your intuition.

Taking it a step further, if pangs of doubt still persist, yet intuitively you know that rapid action is called for if you are to prevent a potentially bad situation from deteriorating further, then tell yourself that nothing is accomplished without shifting to the offensive (see Rule 1). Also, take comfort in the reality that the risk is quite low that your best thought-out decision would easily ruin you, providing you have gathered reliable competitive intelligence, followed the rules of indirect strategy (Rule 2), speed (Rule 3), and concentration (Rule 4). If all your planning is kept within a mask of secrecy, the competition will not see where you are going.

The following case illustrates the far reaching outcomes of competitive intelligence.

Case Example

Intel Corp. made a massive strategy shift in 2006. Almost three decades after introducing its now familiar "Intel Inside" branding logo to highlight its Pentium chip, management decided to replace it. The new design joined the Intel name with the tag line "Leap ahead" to point the company into the future.

This was not change for change's sake or a decision taken casually. Competitive intelligence made abundantly clear the necessity to move beyond microprocessors, as speed alone was no longer a viable differentiator for profitable growth. Another hard reality had emerged: PC growth was slowing, giving way to hot technology products, such as feature-loaded cell phones and various exotic handheld devices. Specifically, intelligence revealed a clear trend toward platforms of microprocessors that combine silicon technology and customized software to power new devices for dedicated applications.

Thus, Intel management faced choices: Stay put, continue to confront grueling competition, live with intense pricing pressure, and search for profitability within a mature market or reach out toward product innovation, growth, and emerging markets. A definitive answer surfaced: Dump the old model of focusing exclusively on PCs. Replace it with a new strategy by redirecting Intel's resources into developing products for a full array of growth fields, including consumer electronics, wireless communications, and health care. The ultimate goal was to provide manufacturers and end users with everything from laptops and entertainment PCs to cell phones and hospital gear with complete packages of chips and software.

Moving forward and relying on competitive and market intelligence to drive its new direction, Intel set in motion a program of sensitizing its personnel to the changing dynamics of consumer needs. The company started hiring a new breed of professionals—sociologists, cultural anthropologists, and even doctors—to develop products working with software developers.

Quick Tips: Questions You Should Ask of Competitive Intelligence

- What is my competitor's overall business strategy, particularly where it impacts my plans?
- What is my competitor's financial picture, including breakdowns of costs and sales by product lines?
- What new products or services are under development?
- What new markets are expected to be targeted?

- How are the competitor's business units staffed and organized, especially in sales and product development?
- What is the caliber of the competitor's leadership?
- What market positions or market shares are held within each product segment?
- Where are the competitor's vulnerabilities by product depth, product quality, customer service, price, distribution, and reputation?

Tools and Techniques of Competitive Intelligence

Different types of tools and techniques are available for competitive intelligence. The following subsections itemize the primary techniques. Select the ones based on the time and resources available to you. In any case, make every effort to use as many techniques as possible.

Sales Force

Begin by gathering relevant data from your front-line sales force. Talk with them. Travel with them. These individuals are in continuing contact with customers and are one of the prime sources of real-time intelligence. To maximize their output, instruct them on what to observe and brief them on the reasons for doing so. This may involve sharing some information about your business plans. Also show them why competitive intelligence is the critical underpinning of all business strategy. A highly useful approach is to require call reports after all customer visits. While product usage, future sales prospects, and similar information are the usual topics reported, require them to add specific information about competitors' dispositions and strengths, significant changes in strategies, and all other relevant activities. Then, have them pass on the reports to an approved list of recipients via e-mail or any other real-time communications channel.

Customer Surveys

Work with your marketing department or an outside market research organization to track competitors' activities. The primary approaches include:

- *In-person interviews:* An interviewer questions individuals in the privacy of the interviewees' homes or offices with the aim of gathering specific information

about competitors. Those interviewed could be users of the product or service, competitors' sales reps, and distributors.
- *Telephone interviews:* These interviews are somewhat difficult to secure with the prevalent use of voice mail and other devices. However, if handled by professionals who can gain the interviewees' attention because of their reputation or with the possible trade-off of offering hard-to-find information, then this technique could be a good source of information.
- *Mail interviews:* Survey questionnaires are useful to acquire product information about competitors' products and services. In such instances, you can avoid using your company name; instead, use the name of a generic-sounding research organization, which could be an actual department of your company. In such research, depending on the type of product, it is useful to offer a financial incentive for taking the time to fill out the questionnaire.

Published Data

Tap into the numerous sources of published information, from small-town newspapers, where even a competitor's minor activities make front-page headlines, to large-city or national newspapers and magazines that provide financial and product information about competitors. Monitor want ads in print and over the Internet for clues to the types of personnel and skills being sought. Look, too, at the growing number of databases that provide voluminous quantities of competitive information. Some of these databases are free and available through public libraries; others are available at affordable costs.

If your competitors are public corporations, refer to their easy-to-access and information-crammed annual reports. With great detail, they reveal (required) financial data, product development plans, and markets they view for future growth. Allied to published data are the speeches by senior executives of competing companies. They offer valuable insights into their firms' future plans, industry trends, and strategies. At times, it is astonishing how much sensitive information is provided in speeches that are given at trade shows and professional meetings.

Government Agencies

Depending on your location, you can access an ample supply of information from government sources through direct access, service organizations, or over the Internet. These sources usually include a vast amount of data from national, regional, and local government offices. Most information is available simply for the asking or at modest cost. In some instances information is available about types of work contracts previously awarded and to whom. Also, competitive information about which companies are currently bidding for new contracts may be obtainable; this provides valuable clues to the type of competitors eligible to bid.

Industry Studies

Management consultancies, universities, trade associations, individual consultants, and investment securities firms conduct industry studies. Also, specialized companies publish industry and competitive information such as:

- Data on a vast variety of products sold online and through retail outlets, television, magazines
- Weekly family purchases of consumer products and information on home food consumption by geographic and demographic breakdown
- Reports of warehouse withdrawals to food stores in selected market areas
- Statistics on television and radio markets according to demographics, experiential life styles, and brand preferences

On-Site Observations

You can handle this form of competitive intelligence personally or under the guidance of trained individuals at competitive intelligence firms. For instance, simply observing numbers of service vehicles and supply trucks in and out of a plant or at store locations provides first-hand information about the level of activity at a competitor's location. In one case, an observer noticed several freight cars at a competitor's railroad siding. However, on closer examination, there was a barely visible layer of rust on the tracks, which indicated little or no recent rail activity. The observer concluded that the freight cars were there as a deliberate deception.

Also, observing the schedules of employees arriving and leaving work can indicate patterns from which you can make reliable assessments. Even watching the movements of individuals in a parking area, the body language of people talking, or the work materials being carried can give clues that further fit an overall pattern. As for a retail store environment, watching and recording individuals entering and exiting, with or without parcels, could provide valuable information when compared to similar locations.

Competitive Benchmarking

Benchmarking is an assessment tool to compare and measure your firm's business processes against those of your competitors. Use it to reexamine your operation and reassess your business strategies and tactics. Also see the discussion on ethnographic techniques in Rule 4: Grow by Concentration. The value of this tool is that you learn how and why some competitors perform with greater success than other organizations inside and outside your industry. You will find such information

Exhibit 5.1 Benefits of Competitive Benchmarking

- Improves your understanding of customers' needs and sensitizes you to the underlying competitive dynamics operating within your industry
- Helps you document why some competitors can perform similar processes at higher performance levels than at your organization
- Creates a sense of urgency to develop long-term objectives and strategies
- Encourages a spirit of competitiveness as managers at all levels recognize that performance levels in best-in-class organizations may exceed their own perceptions of what constitutes industry—and world-class—competitiveness
- Motivates individuals to new heights of innovative thinking and achievement

highly useful when planning strategies and determining which actions are likely to succeed.

Xerox Corporation is an outstanding example of successful benchmarking. The company developed the following 12-step process for competitive benchmarking:

Planning
1. Identify benchmark outputs.
2. Identify best competitors.
3. Determine data collection methods.

Analysis
4. Determine current competitive "gap."
5. Project future performance levels.
6. Develop functional action plans.

Implementation
7. Establish functional goals.
8. Implement specific actions.
9. Monitor results and report progress.
10. Recalibrate benchmarks.
11. Obtain leadership position.
12. Integrate processes fully in business practice.

Exhibit 5.1 summarizes the benefits of benchmarking.

Internal Competitive Intelligence

The preceding techniques deal with the external environment—specifically, the activities of competitors. Now, look internally at your own company's state of competitiveness. To get a complete picture of your organization, evaluate it by using the following approaches*:

- *Performance analysis* relates to your organization's structure and culture; employees' skills, morale, and discipline; internal procedures and systems; state of technology that drives resource utilization, product innovation, and productivity.
- *Strategy analysis* examines your firm's ability to react to aggressive competition, defend existing markets, and attack new markets. It also pertains to employees' capabilities to develop competitive strategies.
- *Strategic priorities analysis* concerns the long-term commitment to quality, customer orientation, and human resource development.
- *Cost analysis* relates to achieving competitive advantage and sustained profitability, which means looking at all business processes and determining when and where cost-cutting programs should be initiated.
- *Portfolio analysis* reviews markets and products and the strengths and gaps within each market segment.
- *Financial resource analysis* studies the availability of cash to support different competitive scenarios, which include initiatives to preserve product quality, maintain customer satisfaction, and provide on-going technical and customer service.
- *Strength and weakness analysis* surveys your company's distinctive competencies and types of unique assets. The intent here is to determine sources for developing indirect strategies. This analysis also includes more qualitative evaluations such as managerial leadership, employee attitudes and motivation, and the level of innovation in new product development.

The intent in all of these internal examinations is to safeguard a viable competitive position.

Reverse Engineering

Here is where you gain the cooperation of technical, manufacturing, and financial personnel to break down competitive products to their component parts. Each part is identified on a range of criteria that covers manufacturing processes, utilization of current technologies, quality, performance, reliability, support services, and applications. The product is then analyzed full cycle, including projections of costs,

---

* There is an increasing use of dashboards in many companies. As information systems become more sophisticated, the ability to have an accurate collection of key performance indicators on a single page is now possible for most small and midsize companies. Dashboards can serve as the primary interface for measuring the success of an organization's strategy as part of a business performance management (BPM) system.

profits, return on investment, and any other financial measures required by your company. A side benefit of reverse engineering is that personnel representing most functions work together with the full awareness of customers' needs and the necessity to maintain a competitive lead. Reverse engineering works well with competitive benchmarking.

Market Signals

The competitive marketplace pulsates with the dynamics of competitive actions. When you tune in to market signals, you gain invaluable insights to feed your own strategy process. For instance, consider the following types of signals:

- Competitors abruptly announce a new value-added service. Equally important, the news triggers interest among your customers.
- Competitors introduce rich financial incentives for distributors to push their products aggressively. Your customers show strong signs of responding to the incentives.
- Unforeseen promotional bursts from competitors siphon off sales you counted on.
- A competitor suddenly adds sales and service reps to a market segment that you thought was secure.
- An enhanced product quietly and suddenly introduced to your customers by a competitor begins to stir interest.

There are other jarring signals that should alert you: Competitors open or close regional offices. Sudden management changes are announced. Upsetting rumors persist of new competitive alliances. Also not to be overlooked are the telltale signs of internal disorder and inept leadership as you observe competitors' personnel. For instance, managers openly show discouragement, display low morale, or exhibit short tempers. Sales reps overtly look for other jobs, gripe about working conditions, or complain about shortages of sales aids and supplies. They mutter about ineffectual leadership and cuts in wages. They whine about stringent rules restricting travel-related expenses or object to executives condoning (or overlooking) abuses in corporate procedures.

At the same time, you may observe lurking signs of general disorder, sloppiness, or indications of internal desperation. These signals could represent changes in a competitor's traditional operating style, or patterns in handling customer relationships or in the general demeanor of how executives interact with their personnel. Also, look at the signals coming from customers who openly complain about a competitor's policies, rules, and procedures. If pumped for detailed information, they often surge forward with a torrent of grievances.

All these words and actions could indicate fear, uncertainty, insecurity, and a variety of deeply seated internal problems. It is in such a state of unbalance and discontinuity that you can shape your own strategy. Therefore, maintain an open, two-way communications channel where signals from sales reps to senior-level management are red-flagged for priority handling.

One other area, usually apart from the general tools and techniques of competitive intelligence and worthy of a separate section, is the human side of competitive intelligence. More specifically, this entails employing *agents*.

Employing Agents

Agents are your eyes and ears at conferences, trade shows, and even at your competitors' locations. They go beyond the raw numbers, charts, surveys, benchmarking, and the other CI techniques. They explore the human side of competitive intelligence by reporting on the behaviors and personalities of key individuals. Their primary tool to dig for information is personal interaction and observation. Agents also screen and interpret events and news, and validate or dismiss information gathered by other means. Before moving forward and employing agents, however, observe a few general cautions.

First and foremost, make certain that you are not violating ethical and legal guidelines and check if you are adhering to your company's policies. Second, assess prospective agents' motivations, personality traits, and talents. Then, you can determine in what capacity to employ them. For instance, some individuals are only interested in money, with minor interest in obtaining accurate information about the competitor's true situation. In such cases, question their integrity and use great care in using them. Third, develop a clear idea about the information you seek. Then make certain the individual understands what you want. The following subsections represent categories of agents, along with suggestions on how to use them.

Native Agents

Native agents are the types of individuals with whom you would normally interact during professional gatherings. They tend to share company information voluntarily to satisfy their personal interests, such as making new industry contacts and advancing their careers. Often, they are somewhat uncaring about their respective company's security or they are simply not cautioned about the dangers of revealing company secrets.

You will find native agents in a variety of places. Trade shows serve as fertile venues for gathering intelligence from those individuals. These are also places where competitors typically reveal extensive information through elaborate demonstrations about their products and freely distribute literature overflowing with facts about pricing, backup services, logistics, product specifications, and so on. Also, key individuals from competitors' organizations often present technical papers at open meetings, which detail sensitive information about upcoming products, services, and even market-entry plans. The Q&A period usually follows, where the speaker, trying further to impress an audience, pours out more information.

Another prime area for intelligence gathering is the familiar hospitality suite at trade shows and professional meetings where alcohol and talk flow freely. It is a spot where security is often lax and everyone's guard is down. College classrooms represent another source of information, particularly where adjunct professors and instructors are also executives and use their respective companies as case examples. In some instances, where fellow students work in companies that interest you, useful data are often revealed about their companies through class presentations, term papers, and casual conversations.

Inside Agents

As with native agents, inside agents work for competitors. In many cases, they may have been bypassed for promotion, feel underpaid and underappreciated, have been relegated to an insignificant job, or generally have been pushed aside in a variety of political or power struggles within the organization. They feel abused and see their careers languishing unless they make some bold move. They may also find themselves surrendering to financial pressures to keep family and self whole; their attitude may be "now or never." You need to assess such individuals carefully for their stability and determine how to use them judiciously. Obviously, you want their information, within the bounds of ethical and legal guidelines.

Beyond personal observations, you would employ inside agents for their expertise to sort out meaningful information from scientific and professional journals, industry studies, or from innovative projects described in articles and professional papers written by the competitors' employees. Product literature and product specification sheets readily available at trade shows and meetings are packed with tremendous detail. Your agents should be able to interpret the data for meaningful intelligence. In-house company newsletters and news releases contain a fountain of information about individuals who left a competitor's employment and may have moved to the consulting circuit. If approached, these former employees may be willing to reveal information, unless specific contractual restrictions apply.

Press releases may include new employee announcements along with job descriptions, contracts and awards received, training programs available, office or factory openings or closures, as well as specific news that reveals competitor's activities. Here, again, your inside agent could handily provide useful interpretations. Also, there is the continuing flow of rumors from customers and suppliers that your agent can sort out and verify. Additionally, there are local sources worth tapping, such as banks, local trucking companies, and real estate offices.

Double Agents

These agents try to extract intelligence about your company. Stay alert to their intentions. Once identified, you could attempt to turn them around and get them working on your behalf. They would then serve in the same capacity as inside agents. Here, too, you can assume double agents seek lavish rewards and may even show similar personality traits and motivations as inside agents. However, it is in your best interest to exercise caution. That is, determine the veracity of these individuals, the reliability of their information, and how long you can expect them to remain loyal to your cause. Once again, make certain you are not violating ethical, legal, or policy guidelines.

Expendable Agents

These agents are your own people who are deliberately fed inaccurate information, which is disseminated in a variety of ways to cause competitors to make wrong decisions. These contrived leaks take many forms—for example, passing fabricated information about new product features through sales reps who come in contact with competitors' reps, or product managers revealing false dates about a product launch that would disrupt a competitor's plans. In spite of your possible discomfort when undertaking such activities, look at the situation from strictly a strategist's viewpoint. Misinformation needs distribution to divert competitors from directly opposing your *indirect* strategy moves. You thereby preserve your company's hard-won market position, control needless expenditures of financial and human resources fighting unnecessary market battles, and avoid disrupting your strategies.

Living Agents

These agents usually provide the most credible information. They are generally experienced, talented, and loyal individuals who can gain access to, and become intimate with, a competitor's high-level executives. They sit in a position to learn their plans and observe movements. These individuals are truly the eyes and ears and often enjoy the closest and most confidential relationships.

Perhaps the one unsettling issue to cope with when using agents is which of your employees knowingly or inadvertently passes on your company's information directly or indirectly to competitors. Eventually, those individuals are exposed and you can obtain valuable clues about what motivated them to those acts. Another concern is that engaging in such stealth activities is usually contrary to the types of practices most managers care to undertake. Again, think of business intelligence as essential to running a company in a highly competitive environment. Above

> **Exhibit 5.2 The Competitive Intelligence Six-Step Process**
>
> 1. Competitors' size: Categorize by market share, growth rate, and profitability, as well as any other quantitative measures meaningful to your company and industry.
> 2. Competitors' objectives: Determine intentions related to product innovation, market leadership, global reach, regional distribution, and similar areas that would indicate a strategic direction.
> 3. Competitors' strategies: Analyze their internal strategies (speed of product development, manufacturing capabilities, delivery, marketing expertise) and their external strategies (supply chain network, field support, market coverage, and aggressiveness in defending or building market share.)
> 4. Competitors' organization: Examine by organizational design, culture, operating systems, internal communications, leadership, and overall caliber of personnel.
> 5. Competitors' cost structure: Check up on how efficiently they can compete, how long they can sustain pricing competition, the cost or difficulty of exiting a market, and their viewpoints about short-term versus long-term profitability.
> 6. Competitors' strengths and weaknesses: Identify areas vulnerable to a concentrated effort, as well as those strong areas that should be avoided.

all, it is indispensable to the development and integrity of competitive strategies. Exhibit 5.2 summarizes the competitive intelligence process.

Competitive Intelligence Applications

From a strategy viewpoint, you can apply competitive intelligence in the following areas: market selection, maneuvering, and positioning.

Market Selection

Beyond the standard segmentation groupings by demographic, geographic, and psychographic (lifestyle) criteria and the more comprehensive eight categories of markets detailed in Rule 4, use CI to highlight viable opportunities in markets with distinctive characteristics. For instance:

- Look for opportunities where you enjoy superior skills and resources to take a commanding lead in a market segment. That means latching on to new market and product trends (see **Intel** case), adapting online marketing methods to new buying patterns, or finding ways to secure a solid position in the supply chain. It could also mean locating new markets labeled as experiential

markets, where prospects seek highly personalized experiences from using your product or service.
- Similarly to experiential markets, look for a marketplace that has a unique personality where cultural values, histories, and experiences correspond with the individuality of your company's products and services. Included here are markets that cater to individuals sensitive to such issues as the environment, weather, safety, pollution, government legislation, social movements, poverty, human rights, public opinion, workers' concerns, and economic matters.
- Search for segments you possibly overlooked or those that are poorly served. Competitive intelligence is valuable in helping you plug those segments before competitors can move in with their offerings and make your position untenable. In the normal course of keeping CI working for you, new opportunities can crop up at any time as you view the physical and psychological posture of a competitor. Therefore, internalize the basic tenet that marketplaces move and shift relentlessly and are subject to a multiplicity of dynamic forces.

Maneuvering

Maneuvering requires roundabout moves and secrecy. It means taking actions opposite of what your competitor might expect. Doing so avoids showing your hand to vigilant competitors. You also steer clear of direct confrontations, such as damaging price wars. This is why you should rely on CI to make careful estimates about your customers' buying patterns as well as competitors' vulnerabilities. You thereby form more accurate estimates about those two-dimensional conditions that continually face you.

Yet, even with good intelligence, there are still the ever present uncertainties when launching a new product or entering a previously unserved market. If you just repeat yesterday's strategies, then you just blunder ahead chasing an elusive advantage with uncertain outcomes. Instead, incorporate into your thinking and planning a three-part process of *probe, penetrate,* and *exploit.*

Probe means making forays into test markets based on the competitive intelligence guidelines already indicated. First, deliberately probe for viable market segments and obtain feedback about consumer acceptance of your product or service, as well as possible responses to various pricing, distribution, and promotion strategies. You would use the probes to refine your strategies with a series of small-scale market tests in preparation for a fine-tuned rollout. Second, withhold market tests and rapidly mount a full-scale campaign with the primary aim of maintaining secrecy and achieving surprise. Here, again, the aim is to unbalance your competitors into making untimely or rash decisions.

Penetrate means expanding your presence in those segments initially identified in the market probes. Once you have located likely points for market entry, maneuvering takes over to avoid an outright confrontation with any opposing competitor. Here is where you maximize your presence by locking up key customers, securing a workable position in the supply chain, and enriching your product and service offerings to satisfy the specific needs of customers.

Exploit suggests further maneuvering to obtain maximum advantage and increase market share through such strategies as differentiation and value-added services built around the marketing mix.

Positioning

Pioneered by Al Reis and Jack Trout during the 1980s, positioning was popularized as "not what you do to a product. Positioning is what you do to the mind of the prospect." Professor Philip Kotler (Northwestern University) said, "Positioning is the act of designing the company offer and image so that it occupies a distinct and valued place in the target customers' minds." Both definitions presuppose that a product's position results from securing a valued place early on in the mind of the prospect and, barring unforeseen circumstances, retaining that dominant position.

If positioning results from securing a valued place in the marketplace and in the prospect's mind, it comes about as the result of four influences, all of which are triggered by CI:

1. Determine the image of your company, brand, or individual product, from which you can configure strategies and establish a growth-oriented market position (see Intel case).
2. Sort out external market factors and understand the demographic and behavioral characteristics of your target segment.
3. Measure customers' actual perceptions of your products, services, and company.
4. Acquire accurate information about your competitors, as discussed earlier.

The following case illustrates several of the points encompassing the rule to prioritize competitive intelligence:

Case Example

Advanced Micro Devices (AMD), the maker of microprocessors for computers, faced the daunting problem of trying to outmaneuver competing rivals for a viable market position. In particular, competitive intelligence documented that the leading competitor, Intel, had a trusted name in the computer industry, more marketing money, and superior product support. Developing a maneuvering strategy, therefore, meant probing for a soft spot in the rival's market position through which AMD could enter and that it could exploit. Making extensive use of

CI, AMD uncovered valuable information that presented the company with an opening that might crack Intel's hold on the market.

During a two-year period, Intel had been caught napping in designing specialized chips that were both powerful and energy efficient. Its engineers instead worked on the high-end Itanium server chip. Intelligence further revealed that the AMD technology offered large companies substantial savings on electricity and that the chips would take up less space in data centers because they did not require large cooling fans. The situation placed Intel in a position to switch gears and play catch-up. It was during this time-sensitive gap that CI uncovered the opportunity whereby AMD maneuvered rapidly to grasp a meaningful foothold in the market. The strategy unfolded as follows.

First, at the retail level, rather than fight head-on with expensive marketing campaigns, AMD channeled the millions it had earmarked for a branding campaign to consumer rebates, in-store promotions, employee training, and ad space in Sunday circulars. Second, AMD personnel secured partnerships with **Hewlett-Packard**, **Gateway**, **Toshiba**, and **Dell** to use its high-performing chips. Also during that time period, intelligence uncovered clues that Intel was experiencing a parts shortage for desktop PCs. Again, moving rapidly, AMD dispatched top salespeople to fill the supply void with its own chips. That quick move proved successful as AMD-based computers took over larger areas of shelf space at **Best Buy**, **Circuit City**, and other high-profile electronic retail chains. The bottom line is that AMD's worldwide share of mainstream microprocessors rocketed from 5.7 percent to 15.3 percent.

Summary

From the opening moments of a product launch, an effort to dislodge a competitor from a market segment, or a campaign to outmaneuver a market leader, everything is uncertain except what you intuitively sense and factually understand from competitive intelligence. That combination is the only means you have for penetrating the fog that clouds competitive actions. It is all you have to decide how to manage your resources, direct your personnel, and activate your strategies. While your instinctive feelings of what constitutes proper action are always present, what needs your continuous attention is a commitment to use every conceivable approach to obtain reliable competitive intelligence. You are then in the optimum situation to deal with the spontaneous market situations that require your best judgment at the moment.

Before moving on, review Rule 5 using the strategy diagnostic tool to assess how this rule would affect your strategy.

* * *

Strategy Diagnostic Tool

Strategy Rule 5: Prioritize Competitive Intelligence

Part 1: Indications That Strategy Rule 5 Functions Effectively (Contributes to Implementing a Successful Competitive Strategy)

1. After each major competitive encounter, an evaluation meeting takes place that includes comparing competitors' performance against yours, assessing strategies and tactics that worked or failed, and pinpointing most useful competitor intelligence.

   ☐ Frequently ☐ Occasionally ☐ Rarely

2. Management provides adequate funding for competitor intelligence and makes it a key part of the strategy development process.

   ☐ Frequently ☐ Occasionally ☐ Rarely

3. Managers use competitive intelligence as an evaluation tool for assessing the levels of risk associated with various strategy options.

   ☐ Frequently ☐ Occasionally ☐ Rarely

4. Managers use competitive benchmarking to determine areas of vulnerability compared with those of competitors.

   ☐ Frequently ☐ Occasionally ☐ Rarely

5. Managers use competitive intelligence to increase their accuracy in selecting markets, locating an optimum position, and determining how to defend against competitive intrusion.

   ☐ Frequently ☐ Occasionally ☐ Rarely

6. Managers recognize that competitive intelligence allows for preempting and outmaneuvering rivals, thereby preserving resources.

   ☐ Frequently ☐ Occasionally ☐ Rarely

Part 2: Symptoms That Strategy Rule 5 Is Functioning Ineffectively (Detrimental to Implementing a Successful Competitive Strategy)

1. Managers frequently show surprise at competitors' actions; this hampers their ability to respond correctly and rapidly.

   ☐ Frequently ☐ Occasionally ☐ Rarely

2. Managers tend to be indecisive due to lack of real-time information about market events; this filters down and affects the attitudes and morale of employees.

   ☐ Frequently ☐ Occasionally ☐ Rarely

3. Personnel are inclined to exaggerate or underestimate a competitor's situation without any substantive backup intelligence.

   ☐ Frequently   ☐ Occasionally   ☐ Rarely

4. There is a tendency to develop product launch plans without documenting market conditions, competitors' strengths and weaknesses, and buyers' specific needs.

   ☐ Frequently   ☐ Occasionally   ☐ Rarely

5. No workable two-way communications channel exists from the field to executives that red-flags opportunities or threats.

   ☐ Frequently   ☐ Occasionally   ☐ Rarely

The ratings for Parts 1 and 2 are qualitative assessments of managers' overall ability to prioritize competitive intelligence and execute an effective competitive strategy. Based on a diagnosis of your company's situation, use the following remedies to implement corrective action:

- Make competitive intelligence an integral part of the business planning process; use it as the centerpiece for developing offensive and defensive strategies.
- Train staff from various functions, in particular the sales force, to recognize the changing dynamics of consumer needs and competitive threats and the urgency to actively gather reliable intelligence.
- Use a team approach to keep track of competitors' activities, with particular attention to decoding the competitors' strategies.
- Set up a procedure for competitive benchmarking to compare your firm's systems and processes against those of your competitors.
- Incorporate the use of agents to expand the reach of your competitive intelligence activities.

Remedies and actions: _____

_____

Finally, to demonstrate the far-ranging impact of this rule, the following company problems linked to prioritizing competitive intelligence come from a survey of chief executives of medium and large organizations (company names withheld to maintain confidentiality):

"Sluggish market, ferocious competition, and our management's own ineptness pummeled the organization."

"Upstart clone makers beat us with low-cost products through new channels, such as the Internet, phone, and superstores."

"We need to keep the company from becoming an also-ran in the industry. But we don't have enough quality information to guide us."

Strategy Rule 6

Align Competitive Strategy with Your Corporate Culture: The Lifeline to Your Organization's Future

Chapter Objectives

1. Reenergize your corporate culture and outperform your business plan.
2. Use corporate culture as an additional management tool to decode the inner workings of your competition.
3. Identify the characteristics of high-performing business cultures.
4. Employ the strategy diagnostic tool to assess Rule 6.

Introduction

Corporate culture is the operating system and nerve center of your organization. Culture shapes how your employees think and how they react when entangled in a variety of hot spots. Some of these are triggered by internal conditions; in others, the company is threatened by overly aggressive rivals or blocked from securing a foothold in a new market.

Corporate culture consists of the deeply rooted traditions, values, beliefs, and history that power the organization and drive individuals' actions. It is dynamic. It is never static. It forms the backbone of your business strategy. Accordingly, there is no generically good culture as there is no one-size-fits-all competitive strategy. Yet, there are seven universal attributes that sustain a competitively healthy corporate culture, whereby individuals

1. Respond rapidly to changing market conditions
2. Exploit fresh opportunities with a bold and unified approach
3. Strengthen customer relationships as an ongoing corporate imperative
4. Create innovative products and services based on a blend of creativity and technology
5. Adhere to the rules of competitive strategy*
6. React to their leadership with high morale
7. Nurture their personal and professional growth

See Exhibit 6.1 for additional characteristics of a high-performing corporate culture.

In contrast, a staid and uninspiring corporate culture closes its eyes to global competition, vacillates over the competitive impact of new technologies, focuses only on building market share in existing markets rather than pushing the boundaries of new markets, and is generally passive to changing buying trends. Additional negative signs include a conscious disregard for making its corporate culture compatible with a constantly changing marketplace. The result is that the organization loses its edge, complacency spreads, customer focus declines, and originality dries up.

Consequently, a positive, supportive corporate culture drives ambitious business decisions, generates customer loyalty, and ignites employee involvement. You can count on five additional attributes to sustain a supportive corporate culture: *fairness, openness, independence, resilience,* and *entrepreneurship.* These elements, then, become the winning performance combination for a forward-looking corporate culture and the platform for winning leadership (refer to Exhibit 6.1).

Further, for you to operate successfully in a constructive corporate culture you need to tune in to your firm's basic beliefs, moral codes, traditions, and standards. This approach will work for you as long as the underpinnings of your organization's or business unit's culture harmonize with the current competitive environment and your firm's overall strategic goals.

Aligning your business strategies with your corporate culture is precisely what will give your plan the singular quality of uniqueness. In practical day-to-day terms, synchronizing strategies with culture will directly impact the markets you focus on, the image you project in the marketplace, the products and services you deliver, and the competitors you are able to face successfully.

---

* The rules of competitive strategy refer to those described in previous chapters.

Exhibit 6.1 Characteristics of High-Performing Corporate Cultures

Beliefs and Values	Employee Treatment and Expectations	Organizational Structure	Leadership and Business Strategy	Vision and Managerial Competence
Total commitment to customer satisfaction	Fairness in enforcing discipline	Use of cross-functional teams to cultivate originality and innovation	Managers show skill in developing and implementing competitive strategies and tactics	Expertise in business planning
Openness to new ideas	Independence with an entrepreneurial outlook	Rapid internal communications through a flat organization	Maximum use of competitive intelligence	Vision to locate new and evolving market opportunities
Tolerance for employee diversity	Flexibility in encouraging personal growth	Speed of decision making	Alignment of culture with a growth strategy	Ability to nurture the health and vibrancy of the organization
Respect for individual achievement	Continuous learning	Mutually supportive internal relationships	Sensitivity to market changes	Aptitude to conceptualize and communicate a strategic direction
Ethical conduct	Motivated employees	Increased authority and responsibility at the field level	Focus on customer retention	Facility to convert vision to action
Personal fulfillment	High morale and pride in the organization		Emphasis on competent leadership	Sensibility and diplomacy in developing and maintaining cooperative external relationships

From a leadership and managerial viewpoint, if you know how to identify the power of your corporate culture, you will get an unmistakable signal whether your strategies can work under adverse competitive conditions. It also provides useful insights about the success or failure of your entire business plan.

The preceding points are summarized as follows:

- If you are a senior executive or business owner who consciously integrates your business strategy with the organization's culture, you are more likely to succeed.
- If you are a middle manager at the division, department, or product-line level who knows how to write a business plan that builds on the subculture of your business unit, you are more likely to win.
- If you do not internalize how the culture of the organization interweaves with today's hotly contested markets and do not know how to align your strategies accordingly, the results can prove fatal.
- If you recognize that corporate culture envelops your entire organization, such as the caliber of leadership, the vision that drives the business plan, the boldness or timidity of strategies, the commitment to customers' needs and problems, and the care and treatment of employees, then you are more likely to succeed.

Characteristics of High-Performing Business Cultures

Corporate culture, then, is the DNA that is implanted in your strategies. It permeates day-to-day organizational life. It functions as the critical lifeline to your organization's future. It delineates, and delimits, the types and range of strategies you can realistically undertake and provides some measure of assurance that you will realize your objectives.

Corporate culture spans the extreme boundaries of growth or retreat, viability or stagnation, or, in its extreme, survival or bankruptcy. To grasp the underlying nature of your organization's culture and to internalize what makes your organization tick is to foretell whether your plans have a reasonable chance of succeeding. Accordingly, if you take the time to sort through the core values, deep-seated beliefs, and historical traditions that shape your organization's culture, you can control how successful you will be in running your operation.

Such awareness is the primary step in formulating a business strategy. Doing so also strengthens your ability to engage the minds and hearts of the personnel who must take responsibility for its implementation. As a tangible outcome of that effort, you will be able to develop more exacting business strategies and tactics that can win in hotly contested markets. The following case illustrates these points.

Case Example

General Electric Co. CEO Jeffrey Immelt declared, "I'm intense about our competition. But I'm more concerned about our culture and our people." He further defined his concern when he admitted to two fears: first, that GE would become boring and, second, that his top people might act out of fear—that is, some executives would shy away from taking the essential risks to propel the company forward.

From the time Immelt succeeded the legendary Jack Welsh, he pushed for a cultural revival by driving his people to focus on creativity, imaginative marketing, innovation, and risk-taking. This did not mean totally revamping GE's culture. After all, the company became a worldwide leader by adhering to Six Sigma, continual improvement of operations, cutting costs, and making deals. However, Immelt knew that in a slow-growing domestic economy and a volatile global marketplace, he had few alternatives other than to go on the offensive and push boldly into new products, services, and markets. Taking a less risky approach would mean a backward slide from which it would be difficult to recover.

How, then, is a corporate culture overhauled when its beliefs and practices have been the hallmark of excellence and the company remains the envy of executives from other high-profile companies? The following three steps summarize GE's cultural shift:

1. Compensation is linked to new ideas and customer satisfaction, with less emphasis on bottom-line results. It is based on managers' abilities to improve customer service, generate cash growth, and boost sales, instead of simply meeting profitability targets. "Immelt has launched us on a journey to become one of the best sales and marketing companies in the world," says one senior GE executive. Top executives hold phone meetings every month and meet each quarter to discuss growth strategies, think up ways to reach customers, and evaluate new ideas.
2. Executives must go after businesses that extend the boundaries of GE. More than just giving lip service to the order, they must submit at least three "Imagination Breakthrough" proposals per year for evaluation and possible funding. The criteria for submitting the proposals must include taking GE into new lines of businesses, geographic areas, or customer groups. For some executives this approach is somewhat unsettling and counter to the former culture built on nurturing internal efficiencies and generating favorable financial numbers. However difficult the shift in mindset

toward embracing risk is, Immelt supports the effort by fine-tuning all the internal operating systems to make it happen. In other words, the risks are shared and are now culturally indigenous to the firm. To further demonstrate support for the revamped culture, Immelt gave teeth to the business-development effort by selling off numerous less profitable businesses, such as insurance. Also, Immelt followed through by spending heavily on acquisitions in such hot areas as bioscience, cable and film entertainment, security, and wind power—all of which have strong growth prospects.
3. Executives are rotated less often, and more outsiders are brought in as industry experts instead of professional managers. This is a big departure from GE's promote-from-within tradition. Immelt pushes hard for a more global workforce that reflects the markets in which GE operates. He also encourages GE's homegrown managers to become experts in their industries rather than just experts in managing.

While several of the changes at GE seem to deal with current issues related to the economy, in fact, the moves aim at revamping the culture and recasting the company for years to come. That means holding on to the fundamental beliefs, traditions, and values that embody the soul of the company and that still remain intrinsic to growth, as it pushes forward to the next era of growth.

Similarly, depending on your level of authority, you can make minor or major adjustments to your firm's or business unit's culture. The acid test to any changes you undertake or recommend is that they form a logical and workable fit with the objectives and strategies of your business plan. Accordingly, the overall culture of an organization or its individual business units and product lines incarnates the core beliefs and values that drive business decisions, generate customer loyalty, and ignite employee involvement.

Be mindful, however, when you grant permission for others to make business-related decisions that they internalize the intrinsic value of corporate culture and its impact on business practice and strategy. For instance, the internal upheaval leading to the **Enron Corp.** debacle illustrates how its embedded corporate culture, when compared with old economy and new economy behavioral patterns, contributed to faulty decision making and its monumental problems (see Exhibit 6.2). For those rival managers closely observing Enron's business practices and then comparing them to generally accepted industry norms, there could have been clear signals to create opportunities.

What lessons can you learn from these examples? Corporate culture, as the backbone of competitive strategy, builds on seven propositions that will directly impact your performance:

Exhibit 6.2 Business Practice, Strategy, and Corporate Culture

Corporate Culture	Employee Outlook	Remuneration	Leadership	Type of Organization	Business Strategy	Individual Behavior
Old economy	Security	Wages	Top-down	Authoritative	Consistent growth	Conform to regulations
New economy	Employee growth and development	Stock options	Inspired	Team work	Rapid growth	Test the limits of regulations
Enron culture	Personal wealth	Stake in the business	Independent	Self-directed	Unrelenting growth	Do what works

1. Lodged in every organization is a highly individualized corporate culture. This unique organizational DNA has a direct bearing on how you shape strategies to maintain a competitive presence in the marketplace.
2. When challenged by rivals bent on grabbing your market share—and even doing you irreversible harm—implementing counteractions successfully relies on your organization's prevailing culture.
3. An organization's culture governs the range of available options you can use to create value, differentiate a product, and satisfy customers.
4. As a screening device, utilizing cultural assessment tools helps determine if your project is likely to be approved by management (see Exhibit 6.1).
5. Correctly interpreting the deeply rooted values, behaviors, and traditions that drive your organization and inspire your employees can significantly affect the success or failure of your business plan.
6. Even if you are not in a position to alter your organization's culture or business unit's subculture, you should know how to select competitive strategies that will likely gain support within the existing organizational hierarchy.
7. Sensitivity to an organization's beliefs and patterns of behavior acts as both an indicator of past performance and a predictor of what the organization—and you—will achieve in the future.

For these reasons, if you know how to evaluate your competitor's culture as well as your own, you can decisively improve bottom-line performance. Therefore, in all business situations, ask: *What type of corporate culture will it take to do my job?*

The following visionary comments from heads of leading organizations illustrate these points:

> "A business can become stronger by making itself a community of people who share the same ideals and goals, the same corporate culture" (CEO, Disco Corp).
>
> "Sharp has a heritage of creating one-of-a-kind products. It is part of our corporate DNA" (president, Sharp Corp).
>
> "Innovative ideas are born of bold dreams and beliefs, and energized through inspired technology and a clear vision" (president, Matsushita Electric).
>
> "My personal credo of three Cs: Challenge, Create, Commit. I tell all my staff to approach life with a pioneer spirit—several steps ahead of the competition" (president, Itochu Corp).
>
> "We have many challenges ahead of us, but perhaps our biggest challenge is the one we have created for ourselves. I mean growing Toyota into a company that truly matters to our customers, our employees, and to the societies where we live" (CEO, Toyota Motor Corp.).

Where positive values, beliefs, and traditions are ignored, abused, misinterpreted, or carelessly altered, the organization's culture is violated. *Down* comes the instigating manager, which would negatively impact the firm's product line and market position—and even the stability of the entire company. *Up* goes the alert competitor, who sees the disruption and enters the confusion with well-timed actions to fill a market void—providing its culture is tuned in and responsive to such competitive opportunities.

In part, the preceding characteristics are explained in the following case.

Case Example

Sony Corporation had been losing ground in one of its markets to **Apple Computer** with its dominant iPod digital music player. It also felt pressure from upstart electronics manufacturers in China and South Korea. The harsh competition, however, was symptomatic of a longer term trend: Sony had experienced a meager output of innovative products in recent years, although its proud history covering six decades consisted of making smashing breakthroughs in consumer electronics, including one of its most prominent products, the Walkman.

Recognizing the urgency to make bold offensive moves, as well as craft defensive ones to protect its hard-won market position, Sony made a radical departure to assure the company's viability. In 2005, Sir Howard Stringer was selected as Sony's chairman and chief executive, a rare instance of a leading Japanese company turning to a non-Japanese to fill a top position. The Welsh-born executive who ran Sony's U.S. operations is credited with reviving the company's music and movie business in North America. Stringer's job involved integrating the company's formidable prowess in engineering and technology with its powerful presence in entertainment to deliver advanced devices and forms of entertainment to the consumer.

What did the job truly entail? Stringer had to motivate the diverse human and cultural elements within the organization and focus on developing innovative products and programs. Such leadership and motivation had to come from a foreigner who did not know the Japanese language. Stringer's challenge was to engage the deeply seated cultural beliefs of the organization. He had to deal with some employees' steadfast nationalistic feelings and do more than just acknowledge the ingrained attitudes of lifetime workers. Expressed differently, he had to reach the hearts as well as the minds of employees. It is heart that collectively describes the underlying emotional qualities by which he would lead and regenerate Sony. Heart is the pathway to the cultural issues that would create stability out of confusion, enthusiasm from discouragement, and bravery out of fear, for these qualities dominate the hearts and minds of individuals.

Accordingly, Stringer's leadership needed to mesh with Sony's ingrained culture. In turn, his personal leadership style required, among other traits, his ability to project confidence and display the discipline and grit that would prevent employees from caving in on every obstacle or becoming disheartened. Actively reaching out to win employees' hearts meant encouraging them, conveying confidence in their work and attitudes, offering appreciation, and, wherever possible, providing tangible security through compensation as well as through meaningful training.

Stringer had to determine when and how to gently modify the corporate culture that enveloped the traditions, beliefs, and values that historically had driven the company over its 60 years of operations and given it worldwide recognition. Yet, without undertaking those changes, any meaningful strategies regardless of their brilliance would have had questionable chances of success.

What can you learn from the Sony case? How a business strategy evolves and is ultimately implemented generally has the imprint of the individual manager. Yet to succeed, it must incorporate the innate cultural characteristics of the firm. Thus, cultural considerations embrace the human behavior that spans the extremes of courage or weakness, decisiveness or reluctance, ambition or apathy, resolve or indifference. They also embody the hearts and morale of individuals who must live day in and day out with their work and share in the good or bad consequences.

Competitive strategy and human behavior, therefore, are fused tightly to corporate culture with two tangible applications. First, they force you to take a broader strategic view of how to approach markets. They also guide your tactical decisions when launching a new product or service, entering a new market, or expanding an existing market. Second, understanding your organization's foundation culture determines how bold or passive your business plans are likely to be against competitors (see the following quick tips).

Quick Tips: Guidelines to Align Business Strategies with Corporate Culture

Would your corporate culture permit you to:

- Implement strategies bold enough to frustrate your competitor's plans or too limiting to do the job?
- Disrupt your rival's alliances and thereby weaken the impact of the opposing manager's strategies?
- Unbalance the competing manager into making tactical mistakes?

- Spread the seeds of uncertainty and doubt among the competitor's personnel through aggressive actions and rumors, which would result in a dispirited group even before the tough marketing battles begin?

Decoding Competitors

What also follows from the Sony case is for you to become totally familiar with a competitor's culture and its strategies. That means acquiring ongoing business intelligence that can spell the difference between success and failure in any competitive environment. Consequently, it is for you to look at a competitor with a 360° view within a changing marketplace of evolving technologies, environmental concerns, and new customer behaviors. What you are likely to see is a panoramic scene that can impact your decisions about selecting markets, launching new products and services, and devising winning strategies.

That is why sales reps should be an important part of the effort at gathering, analyzing, and acting on business intelligence at the grassroots level, with special attention to competitors' current and likely moves. Doing so will provide additional on-the-ground data about competitive behavior under a variety of competitive conditions. Such information permits you to get a closer look at competitors' innate cultural patterns. It is all part of the decoding process.

The eminent business scholar Professor Philip Kotler* characterizes types of competitors:

- A *laid-back* competitor does not react quickly or strongly to a rival's move.
- A *selective* competitor reacts only to certain types of attacks and not to others.
- A *tiger* competitor reacts swiftly and strongly to any assault on its terrain.
- A *stochastic* competitor does not exhibit a predictable reaction pattern.

The cultural signature of the competitor is a contributing factor in deciding, for instance, when to enter new market segments, if at all. In particular, decoding the cultural character of your rival is a vital factor when confronting a larger and more formidable competitor. It directly impacts which market you select, how and when you confront your competitor, and what considerations you give to the changing dynamics of markets.

---

* See Philip Kotler's excellent book, *Marketing Management,* 11th ed. (New York: Prentice Hall, 2003).

Therefore, combining customers' needs and a competitor's profile into your conscious thinking represents the core ingredients for formulating your business strategy. You thereby are in an optimum position to govern the success or failure of your efforts. Consequently, it is your mandate to familiarize yourself with the inner workings of your organization and that of your primary competitor.

The following examples illustrate the diversity of organizational cultures and their impact on the resulting business strategies. Here, again, keep in mind that corporate culture is the structural backbone of business strategy.

Case Example

The clothing company **Abercrombie & Fitch Co. (A&F)** is best known for its rebellious culture. The company exists by flouting tradition. Chief Executive Officer Mike Jeffries entered the company in 1992 and transformed it from a once staid and very conservative men's haberdashery into a hot popular brand. The change was so successful that by 2005 the company had posted continuous profit gains for 48 quarters—a record unmatched by its peers. Jeffries revived A&F by selling preppy but edgy casual clothing at high prices unheard of in that market.

Where does the corporate culture and business strategy connection appear? Let us examine A&F's CEO for behavioral clues. Jeffries operates A&F on a 300-acre wooded area, where a bonfire burns daily and dance music blares nonstop. Working from a conference room with big windows, rather than from a standard office, Jeffries looks down on the grounds. Instead of formal business attire or even casual business dress, he trots around the building in flip-flops, torn jeans, or shorts. His unorthodox approach carries on to the company's advertising, which is so racy that it drew complaints from parents, which made the clothes even more appealing to kids.

Thus, the 180° behavioral transformation of Abercrombie & Fitch had a resounding effect on reshaping the culture of the organization. It reflected in the personality and mannerisms of the staff, the physical layout of its headquarters, the products produced, and the markets served.

Case Example

In sharp contrast, **Boeing Co.** experienced a somewhat desperate situation. In recent years the defense and aerospace giant had weathered operational snafus, ethical scandals, criminal convictions, and abrupt executive departures. In 2005, in the midst of locating a new CEO, Boeing was characterized as a "dysfunctional corporate culture in need of an overhaul."

The problem at the time was that many likely candidates were offered the job, but declined. The principal difficulty they foresaw in revamping the corporate culture was impatient investors looking for quick fixes and not permitting enough time to make changes. To many savvy executives, that barrier appeared as a lonesome pathway to an end-of-career move. Even if an energetic, inspired, and idea-filled executive were to enter the company, activating any effective strategy would not likely happen unless he or she took the time to address the issue of corporate culture, along with its embedded traditions, beliefs, and history.

Boeing found such a person in M. James McNerney, Jr., who joined the company and began the ongoing and lengthy process of changing the company's culture and moving it forward into the future. This effort is ongoing.

Case Example

Symantec Corp., a security software firm, reflects the personality, drive, and sheer grit of its CEO John W. Thompson. Not shrinking from a challenge, Thompson took on two formidable opponents: influential Wall Street and powerhouse **Microsoft Corp.** The following comment adequately tells about the organization he runs and the operating culture that reflects his personnel, strategies, and general forms of behavior: "I'm not going to let Wall Street dictate or set the strategy for our company; I'm going to set the strategy. I am the most competitive person in the world, and I can't wait to compete against Microsoft."

Within these three examples, one dominant component stands out on corporate culture: *beliefs*.

Beliefs

Beliefs are convictions people hold as true. They are anchored to the attitudes, values, and mindsets that originate at the senior executive level (as indicated in the preceding examples) and filter down through the employee ranks to shape their behavior. Beliefs are also influenced by an individual's upbringing, heritage, religious practice, and community tradition—particularly so in a global economy. In practice, however, those viewpoints reflect in such diverse ways as apathy or support of a new service, originality in designing a unique product, or fear in taking on an aggressive competitor. Beliefs, therefore, can unceremoniously and unconsciously creep into employees' thinking and impact behavior.

Today, as many traditional organizational hierarchies flatten and responsibility filters down to lower levels, most well-run organizations take very seriously the notion that employees are free to feed off their own beliefs. A company's strength comes from such diversity. The obligation, then, is to respect different moral backgrounds and personal convictions, as long as they do not conflict with the core values of the organization.

Your task is not only to recognize that people are different, but also to value them because of their differences with all the potential ingenuity and complexities that reside in their minds. Further, your job is to take advantage of these unique differences and mobilize them into a unified team, as suggested by the Sony case. The pragmatic issue here is that diversity should be viewed as a desirable situation and that people of different backgrounds bring different talents to the table.

Cargill, the agricultural and food-ingredient supplier with more than 100,000 employees in 59 countries, implemented its first diversity blueprint in 1995. The program, known as *Valuing Differences,* focuses on integrating the principles of diversity and inclusion into the company's performance-management system. "Our belief is that diversity is not just about getting people in the door; you must have the right culture and an environment that allows everyone to contribute their fullest to reach their potential. Equal access for promotion and opportunities is a major part of our diversity effort. It is much more than just recruiting," says Cargill's vice president of corporate diversity. The highlights of the Cargill program include:

- The company conducts an annual survey to measure response to its initiative, giving the company a baseline index of how well the diversity initiative works.
- Each business unit writes its annual contract with Cargill's chief executive to outline its tasks and objectives. The contract includes *Valuing Differences* as performance criteria.
- It establishes a mentoring system in which junior-level gay, lesbian, bisexual, and transgender (known collectively as G.L.B.T) employees volunteer to mentor more-senior managers for a year.

Carried to its extreme, diversity is further magnified by the new technologies that can marshal the talents and inputs of millions of people worldwide. For instance, search engine **Google** instantly polls millions of people and businesses whose Web sites link to each other.

Companies from **Procter & Gamble (P&G)** to **LEGO Group** use Internet-powered services to tap into the collective beliefs of employees, customers, and outsiders, which are then used to drive their internal operations and product development activities. P&G now gets 35 percent of new products from outside the company, up from 20 percent three years ago. That has helped boost sales from R&D by 40 percent. LEGO uses the Internet to identify its most enthusiastic customers to help it design and market more effectively. After a new locomotive kit was shown to just 250 train fans, their word of mouth through the Internet helped the first 100,000 units sell out in ten days with no other marketing.

Consequently, if you take into account the expansive role beliefs play in preparing employees to develop and implement a business strategy, you will enhance your chances of achieving objectives. Employees often act and win over tremendous odds when they are convinced of the ideals and beliefs for which they are working. Therefore, be aware of beliefs and culture in three contexts:

- Look to enhance your sensitivity to the diverse backgrounds of your employees, as well as with relationships within the supply chain.
- Increase your awareness of the active culture that is exhibited in your markets.
- Secure a lifeline to your customers by understanding their customs and traditions, and incorporate the significant ones into new product features at the beginning stages of development.

The following examples illustrate the irrefutable relationship of business strategy and corporate culture.

Case Example

Cisco Systems is characterized by (1) a near-religious convergence on the customer, (2) a total belief in employees as intellectual capital, and (3) an energetic willingness to team up with outsiders to develop active partnerships. This passion for molding such an outside-in focus is credited to the leadership of CEO John T. Chambers, who clearly saw those attributes as a value system to drive all subsequent actions. In turn, other levels of employees recognized such values and used them to shape strategies that conformed to the culture of the organization, thereby increasing the chances of successfully implementing business plans. Further, by accepting those three characteristics as a functional formula, line managers were able to reliably predict the success or failure of their respective strategies—depending, of course, on the competitor's ability to challenge them and on whether they correctly interpreted the competitor's culture. Therefore, your job is to become adept at cracking the competitor's cultural code.

Case Example

Charles Schwab & Co. employees initially resisted taking brokerage orders over the Internet. Senior management assembled nearly 100 managers at the base of San Francisco's Golden Gate Bridge, handed each a jacket imprinted with the slogan, "Crossing the Chasm." The group proceeded to march across the bridge and symbolically walk into the Internet Age. That physical act symbolized the cultural reinvention of the company and the strategy of burning bridges behind it.

Value systems and practices are often wrapped in symbols as a means to initiate and sustain change. The following is what you should know about symbols and rituals to enhance your managerial effectiveness.

Symbols and Rituals

Symbols indicate signs, acts, or objects that signify a special meaning. They serve as a form of communication within a culture. If an organization is to maintain any semblance of uniqueness, originality, and competitiveness, it cannot do so without meaningful symbolic representation because organizations can be seen by employees and outsiders only through their associated symbols. Therefore, if organizations are to remain unique entities, they can be represented symbolically through a dress code, an oath, or a song, as well as the manner of addressing people with certain terms or titles. It carries further to the choice of words that represent boldness or weakness of a business strategy, as well as to the expected behavior when entering or defending a market. What follows, therefore, is that an employee's allegiance to an organization is expressed though symbolism.

Connected to symbols are *rituals*. These consist of traditional or contrived ceremonies (as in Schwab's "crossing the chasm") in which some physical act or expressive behavior dominates otherwise technical or rational actions. People tend to think of organizations as physical units and part of the material world. Yet, the reality is that rituals represent the means by which people are linked to organizations to give them a "human" dimension and thereby a differentiated quality.

Since rituals assume various forms, it is highly useful for you to investigate the meanings, types, and structures of the symbols used in your company's rituals. What also follows is for you to be aware of your employees' beliefs in the effectiveness of the rituals. Where accepted and meaningful, a ritual gives employees confidence, dispels their anxieties, and disciplines their work group.

In some situations, rituals may do nothing more than tighten the relationships that assimilate one business unit with another or they may be used to bring together diverse units to work toward a unified corporate goal. Again, in the Schwab case it was expressed by "crossing the chasm."

You will also find cultural symbols and rituals most useful when viewed as a form of communications through the spoken word. Words define and interpret "what is really going on in the company." By this means, they embody each person's behavior through the social network of relationships that exist among the various individuals who are interacting.

The following examples illustrate the various points associated with symbols and rituals and their resulting impact on initiating cultural change within the organization:

- **Intuit**, the finance-software developer, initiated an internal change by issuing a terse press release stating that the company would deliver its new products over the Web, rather than in its traditional software format. Further, the changes would start immediately, without a phasing-in process. The symbolic shock value of the pronouncement jolted the company forward and triggered a rapid cultural change. For some employees, the change was radical and the shock value associated with the decision tended to create an unbalancing effect. In other cases—especially when effective leadership was present to provide confidence and support—the transition resulted in a favorable outcome.

  The primary issue here was to recognize the makeup of the staff, their longevity with the company, and the environment in which they worked. This included their beliefs about the direction of the organization, the manner in which the strategies are implemented, their respective roles in the plan, and the beliefs in the capabilities of their leaders. The managerial question is: What stability or upheaval effect will the changes have on their equilibrium?
- **GE Aircraft Engineers Division** shifted its engineers into a renovated warehouse that had the look and feel of a high-energy startup. The aim was to redefine an existing culture from a slowly moving, tradition-based establishment to one that emphasized creativity and ingenuity. The symbolism associated with the physical layout formed the basis for change.
- **Heineken Brewery** initiated a cultural change at the tradition-bound 400-year-old beer maker. The aim was to stir employees out of their complacency and push Heineken to break away from its play-it-safe corporate culture. Traditional corporate symbols and rituals were shaken by the bold strategy of making a dozen acquisitions, moving aggressively into seven countries in Eastern Europe, capturing the sought-after 20-something segment, and introducing daring new packaging.

The following lessons surface from the preceding examples:

1. If you are a senior executive or business owner, it is in your best interest to pay close attention to your firm's value systems, as well as employees' deeply rooted beliefs, traditions, and symbols, when determining your organization's future direction.
2. If you are a midlevel manager with bottom-line responsibilities, your everyday job will be easier if you match your business plans for a market segment, product line, or sales territory with the unique culture of your business unit.
3. Employees responsible for implementing strategies must be oriented toward culturally diverse markets, remain flexible, and be adaptable to change. They must tune in to the nuances of the markets and be receptive to their unique value systems, beliefs, and forms of behavior.

Energize Your Company's Culture

To give your business plan a realistic chance of success requires that you give corporate culture the same degree of attention as you would to preparing any value-added strategy. It is corporate culture that gives you and your organization a unique identity, a personality that is a fairly accurate indicator of your probable success. Therefore, if you want to maximize your ability to reach your objectives, understand the prevailing culture that supports your strategies. It is one of the infallible rules of successful competitive strategy.

Avoiding this issue, especially in today's hotly contested markets, can prove fatal. The essential point here is that even if you devise a brilliant strategy, implementing it with any measure of success relies on the state of your organization's culture. For this reason, it is in your best interest to sort through the core values, beliefs, and traditions that shape your organization. You can thereby control how successful you will be in running your operation.

Let us now form a still higher platform for understanding the interrelationships of competitive strategy and corporate culture. In particular, look at the following workable techniques to reenergize your company's (or business unit's) culture.

- *Stay on the offensive.* A good measure of boldness is desirable, depending, of course, on the availability of resources and the level of confidence you show in employees. If the risk succeeds, offer ample rewards; if failure results, avoid damaging repercussions. (Look again at Strategy Rule 1: Shift to the Offensive.)
- *Encourage creativity and innovation.* Seek maximum input from all levels of employees. Try new ideas that could lead to new products, evolving markets, or new businesses. Maintain a cultural sensibility that retains an open mind and avoids the idea-killing phrase, "We've tried that." Also, allow sufficient time for ideas to incubate and hatch into new technologies, products, and services. Develop concrete formats for employees to submit ideas. See the General Electric case in this chapter, with particular attention to a system that requires executives to submit three "breakthrough imagination" proposals every year.
- *Learn to live in a flexible competitive environment.* This is a cultural attribute that is often difficult to embed within an organization and equally difficult to instill in employees. It also needs senior management's full support, especially during this period of such severe market volatility. Support means maintaining an outward display of resolute calmness and unshakable confidence. For some employees, however, extreme change creates an unsettling situation where any perceived upheaval in conditions in or out of the organization is difficult to endure. Still, flexibility is a singular characteristic that must

be maintained. As such, it is an imperative for operating successfully in the Internet age.
- *Act as an aggressive competitor.* This combative mindset helps you discover where your firm has an advantage or is at risk. It indicates strengths and weaknesses in your products, services, logistics, and overall organizational structure. It examines relationships with suppliers, intermediaries, and customers along the entire supply chain. The process exposes strong points and vulnerable areas in technology, manufacturing, human resources, and capital resources. It surveys any other area that might open your firm to competitive attacks or prevent you from taking advantage of ripe opportunities. It also unmasks sensitive information on employee behavior and suggests clues on how to undertake change. Such exposure also sheds light on senior executives who cannot (or choose not to) make determined efforts to take on an aggressive posture. In practical terms, few executives are effective for all seasons, and not all individuals are capable of performing optimally through successive stages of a corporate cycle—startup, growth, maturity, and decline—or even within different cultural environments.
- *Build a solid market position.* The aim is to create a unique market position from which competitors cannot easily dislodge you. To the extent you are able, try to create brand equity and brand recognition. Your managerial efforts should be directed toward mounting a long-term positive image for your firm. Research has indicated that high market share equals high return on investment. Some executives go so far as to advise building market share at any cost (a viewpoint, however, that remains controversial). Others believe that chasing market share no longer guarantees profitability. One point that is not controversial is that customer satisfaction and long-term customer relationships remain the enduring principles.
- *Stay close to evolving technology.* Tune in to what is happening in those technologies that can help transform your business to the new economy models. (Look again at the Cisco case.) Several choices exist: buying a technology, investing in startups, or partnering with a compatible company.

You can employ all or some of these steps to drive cultural change and energize your company. How drastic these changes are depends on the severity of your company's problems. Therefore, you can react to problems as they arise or be proactive by anticipating changes. It is all part of leadership and managerial competence, which relies heavily on estimating your internal and external environments, including competitors, suppliers, and, most of all, your customers. In addition to the preceding criteria, Exhibit 6.3 further defines the components and resulting benefits that energize a healthy corporate culture.

Exhibit 6.3 Energizing a Healthy Corporate Culture

Components	Benefits
Diversity	A company's strength comes from its diversity, where respect prevails for different backgrounds and personal convictions, as long as they do not conflict with the core values of the organization.
Fair treatment of employees	Employees support company efforts as long as equality exists—and when rewards and disapprovals are applied consistently.
Generate pride and enthusiasm	Employee zeal spills over to business partners and customers.
Equal opportunity for employees	A heightened spirit of innovation helps employees achieve their full potential. This leads to team cohesiveness and elevates morale.
Open communications	A channel is provided to pass on beliefs and values and unite personnel toward a vision for the future of the organization.
Respect for employee contributions	This enhances involvement and enthusiasm to work toward common goals and strategies.

Summary

The consensus among many executives is that the purpose of corporate culture is to develop an internal work environment that encourages individuals to perform efficiently. Yet, a corporate culture will be relevant only if is it is aligned with the organization's vision, objectives, and strategies, and the dynamic forces that make up the competitive marketplace. Cisco Systems, cited earlier, embodies such a culture as the organization converges with a near-religious attachment to customers, a total belief in employees as intellectual capital, and an energetic willingness to team up with outsiders to develop active partnerships.

By turning to corporate culture and using it as an additional managerial tool, you derive a significant competitive advantage. From a people viewpoint, you can achieve high employee motivation and increased team cohesiveness. You thereby create an ironclad connection between competitive strategy and corporate culture. Corporate culture, therefore, serves as the linkage by which the strategy rules described in the previous chapters are activated.

Before moving on, review Rule 6 using the strategy diagnostic tool to assess how this rule would affect your strategy.

* * *

Strategy Diagnostic Tool

Strategy Rule 6: Align Competitive Strategy with Your Corporate Culture

Part 1: Indications That Strategy Rule 6 Functions Effectively (Contributes to Implementing a Successful Competitive Strategy)

1. Managers acknowledge that culture is the nerve center of the organization and are alert to what makes the organization tick. They actively tune in to the company's traditions, values, beliefs, and history.

    ☐ Frequently   ☐ Occasionally   ☐ Rarely

2. Managers utilize corporate culture as an additional managerial tool to decode the inner workings of competitors. By that means, they can predict how rivals will react under a variety of market conditions.

    ☐ Frequently   ☐ Occasionally   ☐ Rarely

3. Managers recognize corporate culture as the DNA within all competitive strategies. It serves as one of the primary determinants in selecting strategies and tactics that will likely succeed.

    ☐ Frequently   ☐ Occasionally   ☐ Rarely

4. Managers are aware that corporate culture influences their leadership styles and consequently their ability to implement business plans.

    ☐ Frequently   ☐ Occasionally   ☐ Rarely

5. Managers acknowledge that corporate culture is one of the prime differentiators that give their organization a unique identity among customers and competitors.

    ☐ Frequently   ☐ Occasionally   ☐ Rarely

6. Managers realize that their organization's culture shapes how employees think, act, and are likely to perform in a variety of competitive encounters.

    ☐ Frequently   ☐ Occasionally   ☐ Rarely

Part 2: Symptoms That Strategy Rule 6 Is Functioning Ineffectively (Detrimental to Implementing a Successful Competitive Strategy)

1. Personnel operate within a closed-in and uninspiring culture that prevents them from recognizing the realities of global competition.

    ☐ Frequently   ☐ Occasionally   ☐ Rarely

2. The existing culture is passive so that managers fail to internalize the consequences of falling behind in new technology and its potential for exploiting fresh opportunities.

   ☐ Frequently   ☐ Occasionally   ☐ Rarely

3. The corporate culture is incompatible with the changing dynamics of the marketplace; personnel are inflexible to the shifts in customers' buying behavior.

   ☐ Frequently   ☐ Occasionally   ☐ Rarely

4. Management fails to align business strategies with the corporate culture and thereby places business plans in jeopardy.

   ☐ Frequently   ☐ Occasionally   ☐ Rarely

5. Managers tend to violate the core values, beliefs, and historical traditions that represent the organization's deeply rooted culture.

   ☐ Frequently   ☐ Occasionally   ☐ Rarely

The ratings for Parts 1 and 2 are qualitative assessments of managers' overall ability to tune in to their corporate culture and execute an effective competitive strategy. Based on a diagnosis of your company's situation, use the following remedies to implement corrective action:

- Determine if the existing culture is compatible with the long-term vision and objectives of the organization.
- If a cultural change is required, utilize meaningful symbols, signs, and rituals either to force rapid change or to make a gradual transformation.
- Determine which unique qualities of your organization's culture would define a distinctive identity that can be used as a differentiation strategy.
- Through cross-functional teams, open communications, and a total customer orientation, utilize the diverse talents of your employees to shape a healthy corporate culture.
- Consciously align your business plan with your corporate culture.

Remedies and actions: _____

Finally, to demonstrate the far ranging impact of this rule, the following actual company problems linked to the use of corporate culture come from a survey of chief executives of medium and large organizations (company names withheld to maintain confidentiality):

"An element of arrogance in our company's culture leads some managers to ignore the intense competition."
"We need to energize management and smooth over customers."
"We must meld clashing cultures and heal the rifts."

Strategy Rule 7

Develop Leadership Skills: The Moral Fiber Underlying Business Strategy

Chapter Objectives

1. Adopt the qualities of successful leaders.
2. Personalize your leadership style.
3. Power up your business strategy through effective leadership.
4. Apply the strategy diagnostic tool to assess Rule 7.

Introduction

Leadership is about responsibility, accountability, and achieving objectives. Leaders inspire their people, organize actions, develop strategies, and respond to market and competitive uncertainty with speed and effectiveness. Above all, leaders act to win: to win customers, to win market share, to win a long-term profitable position in a marketplace, and to win a competitive encounter before a rival can do excessive harm. If they lose, their organizations and those they manage suffer.

To function as a leader means influencing people by providing purpose, direction, and motivation, while improving the viability of the organization. Therefore, anyone responsible for supervising people or accomplishing an organizational mission that involves managing people and resources is a leader. Taking this a step further, anyone who influences and motivates people to action or affects their thinking and decision making is a leader. Leadership is not only a function of *position* but also a function of an individual's *role* in the organization.

Consider the following examples where leadership is a crossover of position and role.

Case Example

Verizon Wireless launched its version of a music download service to compete with **Apple Computer's** highly successful iPod and iTunes combination. It took a solid dose of boldness to go against an extraordinarily famous product. It also took the leadership of numerous individuals in varying organizational positions and roles to link vision, purpose, direction, and motivation to their respective staffs.

Throughout the organization, Verizon's product designers, engineers, manufacturing, marketing, finance, and an assortment of senior and midlevel managers displayed their allegiance and dedication to activate the corporate vision. As this multilayered blend of leadership looked to the future, they also worked to the immediate demands of the job. Even with diverse managerial techniques, they demonstrated a common ability to make the objectives clear to those they supervised, followed by a vigorous execution of their business plans.

For Verizon to introduce its new product successfully required flexibility and a mix of leadership styles as different situations arose. Further, no one could be cast as the single leader; each also behaved as a subordinate. All members of the organization worked as part of a team.

What lessons can you take away from the Verizon case? In a market-driven, highly competitive environment, work at developing a personalized, yet flexible, managerial style. Anything else will come across to your personnel as artificial and insincere. This is especially important if you expect them to support the overall organization's vision and goals.

Also, if you rely on only one leadership approach, you suffer the consequences of being rigid and will likely experience difficulty operating in situations where a single style simply does not work—that is, some projects are complex and require different management skills at each stage of development. For instance, projects in the early stages of development, where creative insight and patient testing for performance and quality dominate, require a far different leadership style from that of pumping up a sales force when launching a new product. Similarly, products at various stages of their life cycles—introduction, growth, maturity, decline, and phase out—involve different leadership methods to correspond to the varying market and competitive conditions at each stage.

For these reasons, there is no single leadership style for all occasions. Therefore, model your style to fit your organization's overall objectives. Be certain, too, that it conforms to the individual tasks to be performed by role and function.

There is still another issue that affects leadership: how your personnel feel about the climate within your organization. Climate relates to your employees' perceptions and attitudes about the day-to-day functioning of the organization and their respective units. Climate is allied with corporate culture. Therefore, if you are to develop realistic strategies and implement them successfully, it is in your best interest to define the environment in which you work (also see Rule 6 on corporate culture).

Accordingly, answer the following questions to determine your organization's climate and your role in it. Although you may not be in a position to make changes, at least you can point to the negatives and positives that exist in the everyday workings of the organization and thereby personalize your own leadership style:

- Are priorities and objectives clearly stated and do your personnel generally accept them?
- Is there a system of recognition, rewards, and reprimands? Does it work?
- Do you seek input from subordinates? Do you act on the feedback provided? In particular, do you keep your people informed?
- In the absence of instructions, do individuals reporting to you have authority to make decisions that are consistent with your objectives? Do they take the initiative and act in times of opportunity or emergency?
- Are there signs of excessive tensions among employees or acts of competitive in-fighting in the organization? What are the causes?
- Is your leadership style consistent with your company's values? Is there a working climate of trust? Do other leaders make positive or negative role models?

The following examples illustrate these points.

Case Example

Google, the online search giant, creates a working climate whereby its managers display the outstanding qualities of leadership by motivating their employees to innovate in all aspects of their jobs. Recognizing that inventiveness and innovation are the drivers of organizational success, leadership is dedicated to creating a working culture that encourages fresh ideas.

For instance, Google management gives all engineers one day a week to develop their own pet projects, no matter how far from the company's central mission. If work deadlines get in the way of those free days for as much as a few weeks, they accumulate. Also, the system is so pervasive that anyone at Google can post thoughts about new technologies or businesses on an ideas mailing list, available company wide for inspection and input.

What are the leadership traits that support such behavior? First, respect for the individual forms the basis of Google's leadership. In practice it means recognizing and appreciating the inherent dignity and worth of people. Even where some individuals' ideas will not succeed, their efforts are recognized and respected. This is especially relevant working with culturally diverse personnel with a wide range of ethnic and religious backgrounds. Second, at each level, leaders stand aside and let subordinates do their jobs. They empower their people, give them tasks, delegate the necessary authority, and let them do the work.

The fundamental issue here is that the organization is not going to stop functioning because one leader steps aside. Therefore, central to the job of good leaders is the task of helping subordinates grow and succeed by teaching, coaching, and counseling.

Case Example

BorgWarner specializes in technologies related to fuel economy, vehicle emissions, and stability. It is also an organization that displays effective leadership and operates in tune with the best attributes of a successful twenty-first century enterprise.

For CEO Timothy Manganello, the principal challenge for Borg-Warner is to inspire employees to step up the pace of innovation. The challenge is not just hype. He sets aside $10 million in seed money and solicits ideas at company innovation summits. These events usually result in three or four research projects going forward a year, out of hundreds of ideas submitted. In terms of outlay, the investment represents a mere fraction of what BorgWarner spends on research and development.

Recognizing that technology is a leader's ongoing responsibility, Manganello continually learns how to manage it. Therefore, a central role of leadership at BorgWarner takes the route of harnessing technology to maintain a competitive advantage. Associated with this duo of technology and leadership is the value of a disciplined, cohesive organization that is able to ride out tough times and emerge better than it started.

Riding the waves of uncertainty also means battling the human tendency of procrastination and the temptation to wait for every scrap of information before making a decision to move forward. Your best approach is to act proactively:

- Establish primary and secondary objectives so that, if your main objectives cannot be achieved, you have a fall-back position.
- Develop corresponding strategies and tactics for each of your objectives.
- Set up monitoring systems to red-flag problems; this would permit you to manage unexpected situations with contingency plans.
- Work out an exit plan to salvage an untenable situation.

"A good plan violently executed now is better than a perfect plan next week," declared the legendary General George S. Patton, Jr. By accepting Patton's pragmatic wisdom that decisiveness beats out indecision, you can more readily connect to the rules of shifting to the offensive (Rule 1) and acting with speed (Rule 3). Taking prompt action, however, does not mean impetuous moves whereby you run off with incomplete plans or launch flawed products.

Case Example

Computer Associates (CA) had been battered with a pack of draining problems: products were late, service was inconsistent, executives routinely broke rules, blatant errors kept cropping up in financial records, and major customers threatened to go elsewhere.

But that was yesterday. A new regime entered and began an extensive revamping of the organization to clean up the troubles and reset the company on a steady course to the past glory that had made CA a worldwide leader. Effective leadership took hold and began a turnaround. Executives confronted regulators, appeased customers, made peace with Wall Street, boosted staff morale, and fixed significant parts of the business.

Part of the fix meant discharging unproductive managers, initiating strict accounting rules, and taking vital steps to change CA's image among employees, customers, and investors. It also meant digging deep to the core of the organization and fixing its culture.

What lessons in leadership emerge from the preceding examples? Your best approach is to create an environment of trust and understanding whereby subordinates are encouraged to seize the initiative and act with a sense of purpose and loyalty. For employees, it is no giant leap from thinking their leader is disloyal to the organization to thinking their leader will be disloyal to them as well.

You can achieve excellence as a leader when your people are disciplined and committed to the organization's cultural values. Excellence in leadership, however, does not mean perfection. On the contrary, an excellent leader allows subordinates room to learn from their mistakes as well as from their successes. In such a positive climate, people work to improve and take the risks necessary to learn. A leader who sets a standard of "zero defects, no mistakes" is also saying, "Do not take any chances. Do not try anything you cannot already do perfectly, and do not try anything new."

More lessons emerge: When taking over a troubled organization or business unit, most often you must sort through hard facts, interpret questionable data, and end up using gut-level intuition to arrive at decisions. In all, it takes good judgment to size up a situation quickly, determine priorities, and act. Also, you should be able to consider alternative courses of action and understand the consequences of each move. That includes the ability to assess subordinates and peers for strengths, weaknesses, and a willingness to take action. Embedded in such ability is the self-confidence that you will perform correctly in a tough situation. Self-confidence comes from the inner knowledge that you are competent in your job.

What do your employees absolutely desire? They want a self-confident leader who can accurately assess conditions, know what needs to be done, and demonstrate a capability to take appropriate actions.

Self-Confidence

A self-confident mind is not just capable of strong mental exertions. It is one that in the midst of tackling severe problems can maintain its equilibrium regardless of internal turmoil.

What about *your* self-confidence? To expect that you will remain calm and reserved in every competitive situation is asking for a superhuman effort. The agonizing feeling of failure is not a fabrication or an illusion. It is a conscious realization, for instance, that a competitor may be superior for reasons previously unforeseen, but have become disturbingly clear to you as market conditions unfold. Such conditions could legitimately cause a sudden collapse of all hopes, a breaking down of self-confidence. Instead of working energetically with subordinates to stem the tide, each person fears that his efforts will be useless, hesitates where he should move, and leaves everything to fate.

The essential point is that competitive encounters assume that human weaknesses do exist. These materialize as momentary negative impressions of events that can distract you from your overall objectives. You should put such disruptions into correct perspective immediately, however uncomfortable they may appear. For instance, when you go into a local marketplace be watchful that you do not become sidetracked by an isolated incident. Instead, look at the event and fit it into the framework of the entire market scene. Then test it against the broader business objectives of your business plan.

If your self-confidence begins to falter and you feel your problems weighing on you more heavily, it is your obligation to take whatever steps you can to strengthen your own determination, rekindle enthusiasm, and, most importantly, instill renewed hope among all those you manage. Do so by focusing on the business plan as a whole. Once again, it is vitally important that you retain a broad view of the whole situation. Understand, too, that effective leadership shored-up by well thought out strategies can work as a counterbalance to transitory emotions. This will also help you remain steadfast about implementing the business plan. Also, when you are tempted to doubt the correctness of your decisions, keep your faith in the time-tested leadership qualities and strategy principles that follow.

Qualities of Successful Leaders

Employees enter an organization with their own sets of values, developed and nurtured from childhood through lifetime experiences. In varying degrees, their values are expressed as loyalty, duty, respect, and integrity. However, they can also be empty ideals and may not fully surface if their personal behaviors do not mesh with the organization's values, customs, ethics, rules, and other patterns of behavior. Where an interconnect does take place, the leader, through words, deeds, and everyday practices, is better able to communicate *purpose*, provide *direction*, instill *motivation*, hone *skills*, and deliver *action*—for example:

- *Purpose* gives people a reason to do things.
- *Direction* means prioritizing tasks, assigning responsibility for completing them, and making sure personnel understand the goals. The aim is also to deploy resources for the best outcome.
- *Motivation* inspires personnel to act on their own initiative when they see something that needs to be done—that is, within the overall guidelines of business objectives.
- *Skills* relates to knowledge of people and how to work with them, ability to understand and apply company policies and guidelines to do the job, and technical competence to use the required tools and techniques.
- *Action* means assessing a market and competitive situation, looking for opportunities, developing strategies and tactical plans, and implementing them. Part of action requires setting performance standards, thereby allowing personnel to discover for themselves what happened, why something happened, and how to sustain strengths and improve on weaknesses.

Levels of Leadership

You will find that leadership exists at three levels: *direct, organizational,* and *strategic.*

Direct leadership is face-to-face, first-level leadership. It takes place when subordinates are used to seeing their manager all the time. The leader's direct involvement may range from a small department, business unit, or even a division of an organization. In some instances, direct leadership is practiced from the highest organizational level. Former **General Electric** CEO Jack Welch devoted more than half his time to people issues. With his informal face-to-face leadership style, he gained the steadfast respect of his employees. He acted as if each employee with whom he made contact was a friend and associate.

As Welch casually wandered down aisles to check the stacked products, every employee from clerk to factory worker knew him as *Jack*. His favorite practice was to bypass the corporate hierarchy and communicate directly up, down, and across layers of management. His handwritten notes, sent to everyone from direct reports to hourly workers, presented a lasting and, in some cases, an emotional impression. All these actions were intended to lead, guide, and influence the behavior of individuals throughout GE's complex organization.

Organizational leadership influences larger and more diverse groups. Executives lead indirectly, generally through more levels of subordinates than do direct leaders. The additional levels, however, can make it more difficult to see actual results. For the most part, organizational leaders deal with more complexity, more people, greater uncertainty, and more unintended consequences.

This is especially so when an increasing number of companies are becoming virtual, globally distributed corporations, where many services and functions—from call centers, R&D, and manufacturing to back-office operations—are outsourced. This increasingly prevailing business model means adding a new dimension to organizational leadership from managing the few to the many, regardless of location. Leaders at this level find themselves influencing people more through policy-making and systems integration than through face-to-face contact. They focus on strategic planning and global competition over the longer term.

Yet, staying focused on developing meaningful collaborations, sharing goals, and applying the principles described before remain steadfast leadership rules as leaders learn to leverage global talent. Nonetheless, as practiced by Jack Welch, regularly getting out of the office and visiting outlying sectors of the organization where the daily work takes place is still especially important.

Strategic leadership is prevalent in larger organizations among high-ranking executives. They establish organizational structure, allocate resources, communicate strategic vision, and prepare business units for their future roles. They also work on complex problems in an uncertain environment.

These strategic leaders apply many of the same leadership skills and actions they mastered as direct and organizational leaders. They process information quickly; assess alternatives, often based on incomplete data; make decisions; and generate support. However, their decisions affect more people, commit more resources, and have wider ranging consequences in both space and time than do decisions of organizational and direct leaders. In smaller organizations or in business units of larger ones, all the levels of leadership are usually focused with the owner or general manager.

In a more specialized application, field sales managers at a division of **Hoechst** take on the additional roles of organizational and strategic leadership for their respective sales territories. In addition to their direct leadership role of motivating, training, and coaching salespeople reporting to them, these sales managers hone their skills to shape a strategic vision for their respective territories, develop long- and short-term objectives, design action strategies and tactics, and provide useful feedback to developers for future products and services.

As leaders, they are trained to think through a problem, demonstrate self-discipline, and act with the initiative of a self-starter. That is, they move when there are no clear instructions, act when unexpected opportunities appear, react when the competitive situation changes, and maintain flexibility when the original business plans fall apart. As important, they do not only give orders. They are careful to communicate clearly the intent of the specific objectives to be achieved. All this effort is anchored to *discipline, character,* and *ambition*. These three characteristics do not just appear. They must be developed, practiced, and honed within the leader and among subordinates.

Discipline comprises order, self-control, restraint, obedience, and deference for authority, as well as a sense of duty. It is through discipline that you shape a united effort, which is then fortified through training. The one reality you have to face is that, in day-to-day competitive encounters, your personnel are enveloped by the instinct of self-preservation that takes hold over all other emotions. The aim of discipline, therefore, is to control that instinct. In even the most pressured situation, the individual tends to lose his reasoning power and becomes instinctive.

Only with discipline is there good order, a unity attained from pride. Pride, in turn, exists among individuals working as a team who know each other well and who share the company spirit. Therefore, it is necessary for an organization that seeks unity to create a uniquely individual company culture (see Rule 6). In too many large organizations only the numbers are seen; the individual often disappears. Yet for anything significant to happen, it is the mind and quality of the individual and the unity of team effort anchored to discipline and organizational support that produce any meaningful effect.

The following case provides a pragmatic setting to demonstrate how unity created by team effort resulted in a dramatic turnaround for a company in trouble.

Case Example

Electrolux is the world's number two maker of home appliances. As with many other organizations, large and small, it got caught up in a maelstrom of surging costs, raging competition, and declining market share as cheaper products arrived from Asia and Eastern Europe.

Hurled into action, CEO Hans Straberg initiated a two-phase action plan. Phase 1 dealt with costs as he took the path used by many other organizations hit by low-priced rivals. Straberg closed plants in Western Europe and the United States and shifted production to lower cost locations in Asia and Eastern Europe. It was phase 2, however, that broke new ground. Straberg eliminated barriers among departments and forced his designers, engineers, and marketers to work together as a team to develop new products.

At one brainstorming session, 53 Electrolux employees from various divisions and job functions gathered in Stockholm in search of inspiration for the next batch of hot products. Divided into teams, each group worked on a product concept. When they were finished, a few members of the group dashed off to the machine shop to turn out a prototype. Others remained to draft a marketing plan based on observing actual patterns of product usage among consumers and prospects.

For Electrolux, the dynamic groupthink is a major cultural change. "We never used to create new products together. The designers would come up with something and then tell us to build it," declared a veteran employee. The unified team approach resulted in a doubling of product launches compared with former levels of output. The number of launches with overforecast unit sales ran at 50 percent of all introductions, up from 25 percent previously.

Character describes an individual's inner strengths. It helps the individual know what is right; more than that, it triggers the courage to do what is correct regardless of the circumstances or the consequences. It is character that influences an individual's behavior as it links intuitive insight into action.

However, to hold to a consistent course is difficult. The steady stream of erratic impressions eating at your convictions never seems to cease. Even the greatest stability and determination cannot protect you completely. As already pointed out, negative impressions are strong and vivid and can affect the mind almost unconsciously. It takes strong feelings and strength of character to overcome them. In that regard, there is a generally held viewpoint that executives with deep feelings for their personnel are not suited for managing serious competitive operations. Yet, examples abound to the contrary among such well-managed companies as **3M**, **Microsoft**, **Sony**, **Dell**, **Nokia**, **Starbucks**, and **Samsung**, where there has been a great deal of sentimentality and displays of compassion by leaders for their employees.

At critical times, however, those executives were also able to master their feelings and show dispassion and determination. Deep as their feelings were, their minds kept their poise and equilibrium. Executives with strong feelings and great strength of character, plus a solid dose of good sense, add up to outstanding leaders. These qualities make it possible to see the critical issues in any event. Stubbornness, on the other hand, distorts determination. It borders on vanity, but goes deeper and is displayed as a narrowness of mind, which views giving in on one's own ideas as a sign of weakness.

Ambition is one of the essential qualities of a leader. There never was an outstanding leader without ambition. Ambition is the mainspring of all actions. But for pragmatic meaning, ambition must be worthy of the organization's mission and not a pathway to solitary power.

Ambition is difficult to separate from courage. In analyzing great leaders it is generally impossible to decide which of their actions in the face of severe problems bore the mark of boldness or that of ambition. Both are characteristics of the truly outstanding. It is constructive ambition and the intense desire to excel that stimulate ambition in others. The magic of winning always arouses ambition, which gives momentum to the organization. Therefore, nurturing constructive ambition is another prime duty of the leader.

Yet, the unwelcome reality exists that unrestrained personal ambition does live, with all its excesses and potentially harmful outcomes—as shown by the highly publicized **Enron** debacle. It is through uncontrolled raw ambition that similar scandals have destroyed employees' careers and the economic livelihoods of communities in which the organizations operated.

Embedded throughout any enterprise is the deeply rooted and monumental influence of organizational culture. With its long lasting and more complex set of beliefs, customs, values, and practices, culture is expressed in a corporate climate of how people currently feel about their organization. Corporate culture, therefore, functions as the backbone that supports discipline, highlights character, and encourages positive ambition.

Consequently, if you seek competence in leadership, you must understand the power of corporate culture. It is the cement that binds together all the qualities and gives an organization a unique personality. Expressed another way, corporate culture combines qualities that give solidarity to the organization and form the underpinnings of unity. For these reasons, corporate culture becomes an essential part of your skill set, regardless of your operating levels.

Quick Tips: A Leader's Skills

Interpersonal skills. These affect how you deal with people. Skills include coaching, counseling, motivating, and empowering. It is the ability to communicate your intent effectively, without impatience or anger. If you want your subordinates to be calm and rational under pressure, you must set the standard.

Conceptual skills. These enable you to handle ideas. Such skills require sound judgment, as well as the ability to think creatively, reason analytically, and act ethically. It also means sensitivity to the group's shared set of beliefs, values, and assumptions about what is important.

Technical skills. These are job-related skills you must possess to accomplish all your assigned tasks and functions. Included here is a wide-ranging awareness consisting of (1) respect for the diverse backgrounds of your personnel, (2) responsiveness to the varied cultures of your markets, (3) sensitivity to the customs and traditions of the intermediaries and customers with whom you interact, and (4) insight into the culture and practices of your competitors.

Tactical skills. These apply to solving tactical problems, which include handling localized situations concerning the deployment of resources. Tactical skills combine with interpersonal, conceptual, and technical skills to achieve objectives.

Leadership in the Competitive Arena

As noted in the preceding quick tips, leadership demands, among other attributes, calmness and patience. These are indispensable traits to uphold, yet difficult to sustain when immersed in a competitive struggle. Without exercising rational behavior under pressure, however, even the most brilliant qualities of the mind are wasted. In short, you have to observe what forms of behavior kick in during a tough competitive encounter: fight or flee.

This means that you examine how you react in a variety of situations. Look, too, for behavioral reactions among your employees, particularly in their responses to stress when pushed against hard-nosed competitors. Such understanding is vital and will impact the decisions you make, especially when you face the critical choice between taking the most daring or the most cautious solution.

Making such a choice is totally realistic, as discussed at length in Rule 1: Shift to the Offensive. The essential point is this: The nature of modern day business thinking in a competitive arena requires decisive and bold action. To take bold action entails effective leadership with motivated and disciplined employees ready and willing to implement the strategy.

Case Example

Apple Computer CEO Steve Jobs illustrates such a leader. Within nine years after returning to Apple, the cofounder of the company initiated giant changes in fields of music, movies, and photography. From a company that was nearly trounced by the likes of Microsoft and other aggressive competitors, Apple is now a high-flying innovator in those areas.

What is Jobs's personal leadership style that triggered the turnaround? He used a combination of (1) clever business strategies, (2) absolute management control, and (3) dazzling innovation. Consider the following examples:

Business strategies. As music, movies, and photography went digital, Jobs immersed Apple in developing elegant and simple devices that captured consumers' attention. To create a unique and consistent corporate image, he consistently tried to stand out from rivals by focusing on counterculture themes. Jobs also chose his partners carefully by associating with artists and music groups, such as U2, that are consistent with the corporate impression he tries to burnish in the minds of consumers. His attention to business strategies even goes as far as approving companies for distributing Apple products.

Management control. Jobs insists on personally controlling all aspects of a product—from hardware and software to the services that come with them. Contrary to generally accepted management practice, he micromanages with a total hands-on approach within his area of expertise. Jobs believes in harnessing the dynamism created by small teams of top talent. He also opts to retain them on-site rather than do a wholesale outsourcing of functions overseas with the sole aim of saving money.

Product innovation. Most of Jobs's time is spent coming up with the next blockbuster product to match the astonishing success of iPod and iTunes. To keep on top of the latest technologies, he personally meets not only with suppliers but also with suppliers' suppliers.

What can you learn from the Apple case that relates to leadership in the competitive arena? First and foremost, develop your own *personal* style of leadership. Do so by tailoring it to your character and personality. However, the key here is to be certain that your style harmonizes with your organization's prevailing culture and, of course, with the competitive job at hand.

While leadership relies on basic tenets regarding the treatment of individuals and dedication to clear-cut objectives, it is not an off-the-shelf package created by someone else for you to adopt. That is, leadership is not a one-size-fits-all style. It should vary with the strategy you develop and the type of workforce you manage. (Flexible leadership was highlighted in the previously cited Verizon case and the lessons that followed.) Accordingly, look closely at your own behavior and match it to the attributes mentioned in this chapter.

If your strategy is going to succeed, your subordinates are the ones who must convert it to action. Therefore, take note of your employees' behavior and look out for any of these ominous signs: disorder, complaints, silence, confusion, and any signs of agitation before, during, and after a competitive encounter. Also notice if their feelings are anchored to the fear of losing pride, status, or employment or the effects of losing face in the eyes of peers. Also, look carefully for the social interactions that exist among members of a group. To that point, use a constructive technique of social network analysis.

Social network analysis is a management technique that is likened to a corporate x-ray. It shows how work really gets done among personnel. It helps you trace the invisible, informal connections between people that you tend to miss on a traditional organizational chart. You can do the analysis on paper or, depending on the complexity of the organization, with the use of a software program. You can then plot as a web of interconnecting nodes and lines the relationships among your people and the frequency of contacts.

Looking somewhat like an airline's hub-and-spoke route maps, it helps visualize collaborations and points of contact that you may not be aware of. It is especially useful in understanding the dynamics that exist within the organization or in a group. The technique permits you to find the touch points that create unity and, most importantly, spur innovation (more on social network analysis in Rule 8).

Leadership and Implementing Strategy

An intrinsic part of the preceding case is that your best strategy cannot produce good results if your short-term tactics—those actions that are in the minds and hands of your front-line personnel to implement—are flawed by negative mindsets and behavior. It is at this critical juncture that your confidence as a leader and manager must show through—not just with animated and transparent confidence, but rather with intimate, firm, conscious confidence, which can fire up employees to act with sustained energy and does not disappear at the moment of competitive tension.

It is not difficult to see that individuals move by emotion, strength, and courage in the face of pressure. However, as already noted, if there is a lack of discipline, they are not suitably trained, or solid support is withheld by the organization, your personnel will be overwhelmed by a competitor whose employees may be individually less brave, but as a unified group firmly organized, trained, and disciplined.

Organizational solidarity and confidence cannot be improvised. They are formed of mutual respect and acquaintanceship, which establishes pride and makes for unity. From unity comes the feeling of energy. In turn, it gives to the human effort the courage and confidence of success. Thus, the cliché "all for one and one for all" remains solidly valid.

Courage and unity, then, are the core ingredients that dominate the will over instinct, especially where the stakes get down to winning or losing. Therein lies the justification for forming teams (see Exhibit 7.1).

Exhibit 7.1 The Functions and Responsibilities of a Team

Functions
- Defining the business or product strategic direction, also known as vision and mission
- Analyzing the environmental, industry, customer, and competitor situations
- Developing short- and long-term objectives and strategies
- Preparing product, market, supply chain, and quality plans to implement competitive strategies

Responsibilities
- Creating and recommending new or additional products and services
- Approving all alterations or modifications of a major nature
- Acting as a formal communications channel from the market back to internal departments
- Planning and implementing strategies throughout the product life cycle
- Developing programs to improve market position and profitability
- Identifying market or product opportunities in light of changing consumer buying patterns
- Coordinating efforts with various functions to achieve short- and long-term objectives
- Coordinating efforts for the interdivisional exchanges of new market or product opportunities
- Developing a strategic business plan

Power Up Your Business Strategy

Under exceptional conditions, individuals can accomplish much more than is ordinarily considered feasible. In fact, the more a leader habitually demands from his employees, the more likely he will get what he calls for. Further, within the framework of powering up your business strategy, assess your situation with the following questions:

- How well have our employees been directed during the active part of a campaign or at a particularly difficult competitive event? Were there intervals of confusing and misunderstood directions? Did our organizational chain of command help or hinder good communications from decision-making managers to those in the field who had to implement the directions?
- Were our personnel adequately briefed about the big picture or was it necessary to hold back information or limit the detail about the full business plan?

Was there a point at which certain individuals tended to quit the effort or shift into neutral? In contrast, was there evidence of others moving forward?
- At what point in the control of employees or in overseeing the general market situation did we feel events slipping from our hands? Was there a time when we thought it was no longer possible to recover a position? What signs, signals, or systems could have alerted us to such an outcome and provided time to salvage the situation?
- At what point did control slip from senior management's grasp? Where and when did the break take place? Where and when did management resume control, if at all?

The major lesson from the preceding assessment is that you must use your creative imagination, training, and experience to prepare yourself to work in such a vacuous medium. It is a mode in which your eyes cannot always see, your best deductive powers cannot always fathom, and you rarely can become completely familiar with a jumble of events. This seemingly unsettling condition is absolutely logical, considering the invisible and powerful forces of an internal corporate culture that sometimes clash and at other times mesh. Then, there are the external dynamics of changing consumer behavior and uncertain competitors' actions that crop up at any time.

Your one remedy is to cultivate positive imagination as an essential ingredient of your skill set and an integral part of staff training. The ability to form accurate mental pictures of a situation quickly is especially important to sort out conflicting actions. Conversely, uncontrolled imagination can be disastrous. That is, if you fail to see events as potential opportunities, your mind leads you to form negative conclusions.

The notion of using imagination makes particular sense in light of the tangled markets in which you often operate. In contrast, taking the easier path of repeating yesterday's worn-out strategies and superimposing them on a fresh set of market conditions can only lead to negative financial results, lower employee morale, and the loss of unrecoverable resources. However, imagination is not a fanciful exercise in freefall. Enhance it with all the leadership qualities listed before. Anchor it to the core strategy rules discussed in previous chapters, which have endured in the past and will continue to be viable ones in most competitive situations you will likely face.

In sum, if you train yourself to manage the ongoing pressures of competitive encounters with discipline, with positive imagination to see you through the inevitable fog of uncertainty, and with business plans firmly attached to the strategy rules discussed thus far, you can turn even minor defeats into major victories. In the following chapter, Rule 8 elevates leadership with the additional dimension of morale.

Before moving on, however, review Rule 7 using the strategy diagnostic tool to assess how this rule would affect your strategy.

* * *

Strategy Diagnostic Tool

Strategy Rule 7: Develop Leadership Skills

Part 1: Indications That Strategy Rule 7 Functions Effectively (Contributes to Implementing a Successful Competitive Strategy)

1. Managers score high in interpersonal skills, as well as in their ability to recognize the inherent dignity and worth of those they manage.

   ☐ Frequently ☐ Occasionally ☐ Rarely

2. Leaders demonstrate expertise in developing competitive strategies.

   ☐ Frequently ☐ Occasionally ☐ Rarely

3. Managers show leadership skills in motivating their employees to win: to win customers and to sustain a long-term profitable position in the marketplace.

   ☐ Frequently ☐ Occasionally ☐ Rarely

4. Managers demonstrate superior leadership skills by openly communicating to personnel a clear vision, purpose, and direction for the company and their respective business units.

   ☐ Frequently ☐ Occasionally ☐ Rarely

5. Leaders show superior ability by helping subordinates grow and succeed through ongoing training and coaching.

   ☐ Frequently ☐ Occasionally ☐ Rarely

6. Managers understand that leadership means inspiring their people, organizing actions, developing strategies, and responding to market and competitive issues rapidly and effectively.

   ☐ Frequently ☐ Occasionally ☐ Rarely

7. Managers recognize that there is no single leadership style. Instead, a competitive environment exists that requires a personalized and flexible managerial style.

   ☐ Frequently ☐ Occasionally ☐ Rarely

Part 2: Symptoms That Strategy Rule 7 Is Functioning Ineffectively (Detrimental to Implementing a Successful Competitive Strategy)

1. Personnel mistrust their managers' abilities to assess market and competitive conditions correctly and make timely and accurate decisions.

   ☐ Frequently ☐ Occasionally ☐ Rarely

2. Employees exhibit negative behavior, confusion, and an unwillingness to take the initiative.

  ☐ Frequently ☐ Occasionally ☐ Rarely

3. The organization or business unit is enveloped by a general malaise that creates high levels of stress and anxiety.

  ☐ Frequently ☐ Occasionally ☐ Rarely

4. Managers do not have an effective system of recognition and rewards.

  ☐ Frequently ☐ Occasionally ☐ Rarely

5. Employees show excessive fears of losing: losing pride in the organization and their managers, losing status, losing in the eyes of their peers, and possibly losing employment.

  ☐ Frequently ☐ Occasionally ☐ Rarely

6. Subordinates are rarely asked for input; if it is offered, feedback is seldom given.

  ☐ Frequently ☐ Occasionally ☐ Rarely

7. Overall, leadership styles of senior and midlevel managers are not consistent with the company's core values and accordingly these managers do not make acceptable role models for junior managers.

  ☐ Frequently ☐ Occasionally ☐ Rarely

The ratings for Parts 1 and 2 are qualitative assessments of managers' overall ability to use leadership skills to execute an effective competitive strategy. Based on a diagnosis of your company's situation, use the following remedies to implement corrective action:

- Communicate to staff a vision for the corporation (or business unit) so that they can picture their individual roles in the organization's long-term outlook.
- Permit individuals to act on their own initiative.
- Motivate employees to improve their skills by means of ongoing training and individual coaching.
- Create unity of effort through discipline and fostering a team effort.
- Develop a leadership style that harmonizes with the culture of the organization, objectives to be achieved, and strategies to be employed.

Remedies and actions: _____

_____

Finally, to demonstrate the far ranging impact of this rule, the following company problems linked to the use of leadership skills come from a survey of chief executives of medium and large organizations (company names withheld to maintain confidentiality):

"Worried over complacency. Also concerned over the sharp drop in orders due to the crisis in the industry."

"With the competition circling, it's clear that complacency is a luxury we can't afford."

"Our market is still in the dumper, ensuring yet another loss. There's also a lack of cohesiveness and focus."

Strategy Rule 8

Create a Morale Advantage: Engage Heart, Mind, and Spirit When All Else Fails

Chapter Objectives

1. Integrate morale into your leadership style.
2. Control how morale impacts the outcome of your competitive strategy.
3. Eliminate barriers that inhibit morale.
4. Identify the steps to activate high morale and maintain organizational momentum.
5. Employ the strategy diagnostic tool to assess Rule 8.

Introduction

Morale stands high as one of the most influential forms of behavioral expression. It affects day-to-day employee performance and contributes ultimately to how well a business plan is implemented. It is one of the human dimension's most important and compelling prerequisites to successful functioning. The level of morale is a gauge of how people feel about themselves, their degree of participation in a team effort, and the confidence they show in their leaders. High morale results from effective leadership, mutual interest in solving problems, and widespread respect for peers and managers. Morale forms an emotional bond that springs from common values, such as loyalty to fellow workers and a belief that the organization will care for them. Where morale is at a high level, it results in a cohesive team effort that enthusiastically strives to achieve common goals. Successful leaders know that morale—this vibrant and essential human force—holds a team together and keeps it going in the face of the inevitable problems and reversals.

Morale is triggered through a variety of leadership approaches and organizational combinations. Many are offshoots of two dominant methods: (1) the popular, participative, and "touchy-feely" style that many organizations embrace today, which is diametrically opposite of (2) the autocratic style that is less prevalent among major organizations, but still finds a home in some top firms. Organizational combinations refer to hybrid structures where variations in leadership and management styles, along with organizational designs, surface at different times, depending on market and competitive conditions.

A noteworthy example where the two dominant approaches lived successfully (at different times) is described in the following case.

Case Example

From its beginnings in 1979, **Home Depot Inc.** allowed store managers immense autonomy to make their own decisions and run their own operations. Founders Marcus and Blank installed a decentralized, entrepreneurial business model and used a highly personalized leadership style. While the work was demanding, the company grew in a low-profile, collaborative, and mutually respectful working climate. With its humanistic managerial and leadership methods, the Home Depot chain expanded to become the third largest retailer. It was also the youngest company ever to hit $40 billion in revenue, just 20 years after its founding.

A turning point came in 2000 when a new CEO, Robert Nardelli, took the helm and began a managerial and cultural transformation. Compared to its cultural heritage, the managerial changes were venturesome and audacious. Building on a military organizational model, Nardelli imported ideas, people, and concepts from the military. He initiated sweeping moves to reshape Home Depot into a more centralized organization with a command-and-control management structure. Every major decision and goal at Home Depot flowed down from Nardelli's office. He hired former junior military officers for a two-year training regimen. Overall, some 13 percent of Home Depot's employees have military experience versus four percent at **Wal-Mart** stores.

Nardelli's approach was to build a disciplined corps that would be predisposed to taking orders, could operate in high-pressure environments, and could execute strategies with high standards. In his eyes, it was a necessary step in Home Depot's evolution. Nardelli's spirited methods rekindled stellar financial performance during his tenure: Sales soared from $46 billion in 2000 to $81 billion in 2005, an impressive annual growth rate averaging 12 percent. Profits more than doubled, to $5.8 billion that year.*

Which leadership style is better? Should either be considered a consistent style for all seasons? What benchmarks should be used to determine style: customer satisfaction, degree of competitive aggressiveness, employee performance, stock price (if applicable), gross margins and profits, or high levels of innovation and inventiveness shown by employees? How about the working climate? Should there be a hardball working climate? Should the surroundings match the innate personality of the leader or should they be a flexible work environment that correlates with competitive conditions? Also, what value would you place on the human dimensions of employee attitudes? What are the overall impact of morale, spirit, and the seeming intangible of *heart* on the outcome of your business plan?

What steps should you take to adopt either of the two leadership approaches? What commonalities exist between the two styles—if any? Where Marcus and Blank relied more on instinct than analytics, Nardelli uses a formula of VA = Q × A × E. Its meaning, according to Home Depot, is:

Value-added (VA) of an employee equals the quality (Q) of what you do, multiplied by its acceptance (A) in the company, times how well you execute (E) the task.

For you to find personal answers to the preceding questions and to internalize the contrasting management styles of Home Depot, you will find it highly useful to familiarize yourself (or get reacquainted) with the gurus of motivational behavior: *Herzberg, McGregor, Maslow,* and *Ouchi.* They spawned motivational theories in the twentieth century that are still valid today. As important, you can use them to elevate the morale of your people, with positive impact on implementing your business plans.

---

* Robert Nardelli departed Home Depot in 2007 after a controversy with the board and complaints from vocal shareholders over his extravagant compensation package. According to some industry analysts, Nardelli's performance record excels if measured by his ability to transform Home Depot from a faltering retail chain into an earnings juggernaut.

Exhibit 8.1 Herzberg's Factors Affecting Motivation

Factors Leading to Dissatisfaction	Factors Leading to Satisfaction
Organizational policy	Sense of achievement
Quality of management	Level of recognition
Relationship with boss	Intrinsic value of the job
Working conditions	Level of responsibility
Wages	Opportunities for advancement
Interpersonal relationships	Status provided

Motivational Behavior

Herzberg's Motivation-Hygiene Theory

The most important part of Herzberg's theory is that the main motivating factors are embedded in the satisfaction gained from the job itself. He reasoned that to motivate an individual, a job must be challenging with sufficient scope for enrichment and interest. Motivators—often called satisfiers—are directly concerned with the satisfactions gained from a job. In contrast, a lack of motivators lead to overconcentration on what Herzberg called hygiene factors—or dissatisfiers—that form the basis for complaints. Exhibit 8.1 presents the top six factors causing dissatisfaction and satisfaction.

McGregor's X and Y Theories

These theories remain a valid and predictable means by which you can develop a positive leadership style. It is central to organizational development and corporate culture. McGregor maintained that there are two fundamental approaches to managing people, popularized as theory X and theory Y. Theory X tends to use an authoritarian leadership style. In contrast, theory Y leans toward a participative approach. In today's environment, theory Y generally is accepted as producing better performance, in that it allows people more latitude to grow and develop in self-motivating surroundings. The two styles are contrasted in Exhibit 8.2.

Maslow's Hierarchy of Needs

Maslow viewed people as basically trustworthy, self-protecting, and self-governing. Further, he believed that individuals' innate tendencies are toward growth. According to Maslow's theory, there are five types of needs that must be satisfied before a person can act unselfishly. Figure 8.1 shows these needs, which are arranged in

Exhibit 8.2 Contrasting Views of Theory X and Theory Y

Theory X (Authoritarian Management Style)	Theory Y (Participative Management Style)
The average person dislikes work and will avoid it, if possible.	Effort in work is natural and enjoyable.
People must be forced to work towards organizational objectives.	People will apply self-control and self-direction in the pursuit of organizational objectives, without external control or the threat of punishment.
Individuals prefer to be directed, look to avoid responsibility, generally lack ambition, and want security above all else.	Commitment to objectives is a function of rewards associated with their achievement.
	People usually accept and often seek responsibility.
	People use a high degree of imagination, ingenuity, and creativity to solve business problems.

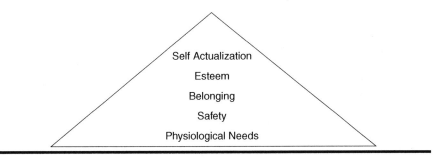

Figure 8.1 Maslow's hierarchy of needs.

hierarchical order and usually shown as a pyramid. The path is to satisfy one set of needs at a time, beginning with the physiological need and then moving upward to self-actualization:

Physiological needs. These cover the basic functions of comfort and maintenance of the body, such as food, drink, heat, shelter, sleep, and health.

Safety needs. These refer not just to physical safety and protection from harm, but also to such areas as financial security, employment, medical and legal assistance, and all means that maintain stability.

Belonging needs. These indicate the need for human contact: family, friends, relationships, teams, and general contact in society.

Esteem needs. These recognize the need for status, power, prestige, acknowledgment, respect, and responsibility. Such requirements provide individuals with a higher position within a group.

Self-actualization needs. After all the previous needs have been satisfied, the top of the pyramid deals with the individual's need to reach for his or her full potential and strive for individual destiny.

Ouchi's Theory Z

This theory is essentially a combination of all that is best about McGregor's theory Y and modern Japanese management. It places a large amount of freedom and trust in workers. It also assumes that they possess strong loyalty and interest in teamwork and the organization. Theory Z places a great deal of reliance on the attitudes and responsibilities of workers, whereas McGregor's X and Y theories are mainly focused on management and motivation from the manager's and organization's perspective.

Now, what conclusions can you make from the earlier questions related to the Home Depot case and the review of classic motivational theories? Draw two immediate conclusions: First, it is possible for both autocratic and participative managerial styles to produce positive results, depending on the level of morale among the participating employees and the leadership skill of the manager. Second, even under a somewhat autocratic, quasimilitaristic work climate, where there is a defined mission, a set of clear and measurable objectives (even when dictated by the CEO), understandable communications, and clearly stated expectations, this style can flourish.* What follows is a set of conditions common to whichever style you adopt:

- Hold fast to the definitive object of all business, which, according to the late management scholar Peter Drucker, is to "create a customer."
- Remove obstacles that deprive people of gaining pride in quality of service and delivering innovative products.
- Break down inhibiting barriers among diverse staffs and create collaborative cross-functional teams.
- Introduce a work environment where the emphasis is on managers leading, not merely supervising.
- Commit to long-term goals, such as attaining market leadership, developing leading-edge products, or maintaining superior service and product quality.
- Eliminate the use of fear as a motivator.

---

* As pointed out in Strategy Rule 7 on leadership: "If you rely on only one leadership style, you suffer the consequences of being rigid and will likely experience difficulty operating in situations where a single style simply does not work."

- Encourage employees to express ideas; listen to them and respond.
- Promote self-improvement as an ongoing imperative. Institute continuing employee training and education to advance their skills, personal growth, and chances for career advancement.

Another commonality that fits into your role as a leader is to heighten morale, which will give you and your staff the resolve and determination to act decisively under competitive pressures. Morale creates cohesion, provides unity, and taps the inner strengths of individuals. Most important, it prevents them from faltering when stress increases due to a tough market situation.

Then there is another reality: Some cavalier executives never seem to worry about morale or even feel harassed. No obstacles seem to bother them. "Here's what I want you to do. Now do it" is a familiar command. Most often, however, they experience incredible disorder in the follow-on moves. They are the great leaders for a day, until the moment that some negative outcome overwhelms them.

The greater and more reliable reality is that achieving high-morale takes ongoing training, excellent leadership, and intimate concern for employees' well-being. In pragmatic terms, marketplace conflicts are generally in the hands of each individual, and never has individual fortitude, triggered by the potent force of morale, had greater importance. This is especially so with the many daily incidents you do not see, as you focus on the more visible competitive conditions that grab your attention.

Morale Meets Technology

Potential problems arise where employee morale confronts technology. Much has been said about latching on to the wonders of technology, even if it means suppressing the human element. Where such a conflict exists, nothing is likely to be more fatal to the organization—meaning: the individual is still the most significant element in conducting business. More precisely, the dominant issue is that people must harmonize with technology to achieve any favorable outcome.

Further, new technologies are almost valueless in the hands of dispirited employees, no matter what their number. It is therefore essential to work at developing their morale, which is often grounded to the organization's embedded culture. Thus, cutting-edge technology and outstanding business strategies, even supported with superior resources, will languish if employees are not personally moved by morale and the driving will to succeed.

The following case provides pragmatic reality to these points.

Case Example

Boeing Co., the aerospace manufacturer, met the enemy and it was itself. Although a high flyer in technology, research, and technical expertise, it was plagued by a dysfunctional culture laced with ethical violations, bitter infighting among business units, and a wave of questionable business dealings characterized by a "win at all costs" mentality. The cumulative effect of these difficulties resulted in Boeing losing its number one commercial aircraft position to Airbus. Also, it was denied access to lucrative government contracts worth billions in potential revenue for several months.

These painful outcomes existed during the tenure of two past CEOs. Then, a turning point began in 2005 with the appointment of a new CEO, W. James McNerney Jr., and his unrelenting moves to revamp Boeing's torpid culture. Recognizing the toxic effects from these deeply seated problems, McNerney immediately set to rebuilding Boeing. The major moves included:

Install a culture built on ethical standards of honesty, integrity, and moral behavior. Previously, executives were silent when colleagues used unethical means to secure business contracts. Under the new guidelines, they are encouraged to speak out against such practices. To give authority to the edict, compensation is tied not only to higher performance but also to higher ethical standards.

Encourage constructive communications. Included here is the former practice of hoarding vital information, especially technical data, and denying access by other business units. Here, too, compliance is tied to managers' compensation. The overall aim is to create a common culture and an allegiance to long-term growth.

Reward managers for meaningful performance. Promotion and increases in compensation are rewards for boosting productivity, as well as driving growth over a three-year cycle.

Use a team approach. The emphasis is on building teams with responsibilities for standardizing and streamlining production lines. The goals are to generate consistent results and lower costs, and improve productivity across all areas of the company, whether on an airplane assembly line or in a research lab. A parallel goal is to end the bitter infighting that prevailed under two previous administrations.

As part of implementing the cultural reformation, McNerney began exerting more effective central control over Boeing's three divisions. Akin to McGregor's theories X and Y, he centralized operations and drew a theory X hard line against unethical practices and demanded fair business dealings. At the same time, McNerney moved ahead with a healthy infusion of theory Y by managing with humanistic leadership. He acted attentively to the small things such as recalling individual's names, listening carefully to their presentations, and not embarrassing subordinates in public. McNerney aimed for a work environment where people talk out when they see something damaging, from sounding off about unethical behavior to stopping a production line if something looks wrong.

Morale Faces the Human Heart

Central to morale and the starting point in all matters pertaining to competitive encounters is to reach the hearts of individuals. "Heart" collectively describes the emotional qualities by which you lead. These include unity, camaraderie, purpose, duty, hope, and the principle factors covered in Maslow's hierarchy of needs. In day-to-day organizational life emotions materialize in conflicting forms, such as through order or confusion, commitment or indifference, boldness or fear, loyalty or deceitfulness. These, too, are the realities that typically dominate the heart.

Not only does heart underlie your role as a manager, it reflects in your outward behavior and ability to perform as an inspiring leader. That is, your demeanor and attitude filter down and impact the performance of staff members who might be low on morale and stripped of courage. Accordingly, you have to show confidence and display the discipline and grit that will prevent you from caving in at every obstacle. Heart is how you create stability out of confusion, enthusiasm from discouragement, and bravery out of cowardice. These, too, are qualities that dominate the heart. On the other hand, if you lose courage, the winning spirit, and decisiveness to complete the planned efforts, then your ability to lead under stressful situations is at serious risk.

Reaching the heart and mind applies equally to competitors. Strategically, your aim is to frustrate the competing manager and move against him. If you aggravate and confuse your competitor into making hasty and unwise decisions, you rob him of the heart to plan with the winning spirit and courage appropriate to the demands of an uncompromising marketplace. As a result, in the hands of an astute manager, this psychological aspect of human behavior becomes an effective strategy tool to unbalance a competitor. Heart, therefore, contains the highly humanistic components that profoundly impact your capacity to manage your people, your competitors, and your ability to implement a business plan.

Tactically, if you move into unfamiliar markets where competitive, economic, and negative consumer influences loom as unexpected barriers, then it is only through the spirited efforts of your people that you can overcome obstacles and push forward. While the inner stirrings of fear and uncertainty can fretfully tug against you, there are also the dominating effects of heart, with all their remarkable qualities, that in the end determine who wins or loses.

In the last analysis, success in encounters is a matter of morale and reaching the hearts of your people. In all matters that pertain to an organization, it is the human heart that reigns supreme at the moment of conflict. Careless managers rarely take it into account. Errors, sometimes irreversible, are the unfortunate result.

Morale Connects with the Power of Unity

You should not permit your employees to fend for themselves. In absolute terms, they must know that support exists at the highest levels of the organization. Also, they must see and feel your physical presence to absorb the psychological comfort and confidence that would sustain their morale and motivate them to keep trying, even under adverse conditions (see Exhibit 8.1).

As for the power of unity, it is not the "Rambo" effect of individual actions that make high-performing employees. It is the collective efforts of individuals within teams interacting among themselves, as well as with those in dissimilar teams, that cultivate unity (see the Johnson & Johnson case discussed later). Therefore, it is your essential role to encourage such interaction. Once again, this is the purpose of teams. (See Rule 7 for duties and responsibilities of a team.)

In the end, it is the manager who knows his or her people best and understands the reasons behind such highly charged displays of behavior as anguish and fear or courage and dogged determination. Therefore, you cannot be a stranger to your employees; it is your influence and physical presence that affect morale. If they feel themselves no longer supported, it creates an uncertain situation for you.

Unity also requires that you sustain confidence in your personal ability to lead. If, on the other hand, you are habitually gripped by fear, then you should give second thoughts about leading your group. Unity is also nurtured by a cohesive corporate culture, which is reinforced by an ethical climate built on core values, as illustrated in the Boeing case. A healthy culture fortifies your leadership role and helps guide your organization or business unit. Consequently, leaders seek to shape their companies' or business units' cultures to support their visions, articulate their goals, and improve their overall performances through high morale.

Quick Tips: Morale and Its Impact on Business Strategy

- Morale suffers and defeats become competitive disasters without sufficient training of leaders and staff.
- Morale fades when there is no inspiring goal to fight for.
- Morale weakens when employees work in a state of uncertainty.

- Poor communications affect morale with a corresponding serious effect on implementing strategies.
- Lack of commitment to a competitive strategy with a determination to win affects morale.
- Unethical behavior impacts morale (see Boeing case).
- An inability to create a long-term vision and develop attainable objectives influences morale.
- Undisciplined behavior that readily gives in to fear affects morale.
- A leader who procrastinates and visibly displays an inability to handle responsibility affects morale.
- A leader who stifles employees' input has a negative influence on morale.
- A leader's inability to accurately assess a competitive situation affects morale.
- A leader who does not know how to reach the heart, mind, and spirit of a group and, instead, shows more interest in his personal agenda impacts morale.

Activating High Morale and Maintaining Momentum

When embroiled in a difficult situation, try to maintain momentum and keep the probability of success on your side. At times a problem may seem insurmountable; nonetheless, you must act against this probability even when the likelihood of success is against you. Do not think of your undertaking as unreasonable or impossible. It is always reasonable and there are always possibilities.

Often, the determining factor for climbing out of a tough predicament and gaining the initiative is to lean heavily on the quality of your personnel. This is possible when you (1) address key issues affecting their morale (see Quick Tips), (2) take initiatives to provide for their growth, and (3) create a feeling of unity within the group. These three areas have as their underpinnings spirited leadership, a clearly understood vision or objective, and competent business strategies.

To turn a potential failure into a sure win also depends on continuing training, communicating purpose, and making certain all the attributes cited previously come into play. The blend of these attributes makes individuals carry on in spite of any inclination to concede defeat.

The following case example relates to these points and the urgency to maintain momentum.

Case Example

Johnson & Johnson (J&J) has been pushing hard to maintain momentum in one of its important and increasingly competitive medical devices business. This blue-chip company has had many notable product successes, such as its drug-coated cypher stent. It also has had its share of failures, such as an artificial spinal disk. On balance, however, there is a clear and urgent need to drive product innovation. The urgency is real: Without a solid jolt in the form of new products and market expansion, earnings could suffer a major nosedive and potentially cede some product leadership positions to competitors.

Managers are driving innovation and creating a sense of urgency in a variety of ways. Organizationally, internal startups have been established. These groups act as if they are totally independent. Some are housed in unusual locations, such as warehouse-like settings where entrepreneurial inspiration tends to surface more rapidly. If any group comes up with a hot idea, they create a business plan and go hunting for financing from J&J's venture-capital arm or from one or more of the firm's existing businesses.

Possibly the most dramatic organizational change has been to establish a centralized, cross-functional position for a chief science and technology officer. The change is noteworthy in that the position exists in an already highly decentralized organization where J&J's numerous businesses operate as autonomous operations.

The officer's primary objective is to find innovative ways to harness the creative energies of individuals in various business units and encourage them to search out ways to combine medical devices and drugs in one product. A secondary objective is to spot opportunities in markets that any one of J&J's existing business units might have overlooked. The new position is also culturally significant in that, in a corporate environment that thrives on autonomy, the change is geared to foster cooperation from dissimilar groups and get them to work together and share thinking, technical expertise, and purpose.

The bottom-line objective is to kick-start the product development effort and maintain momentum for the long pull. The effort is in full swing.

What can you learn from the Johnson & Johnson case? First, overall employee behavior is expressed by an organization's culture, which comes to life through its history, ideals, motives, and deeply rooted beliefs. Second, all organizations remain inherently defective if managers neglect to strengthen the morale of their employees. This means harnessing the creative energies of personnel, whether from product development, manufacturing, marketing, finance, or any other corporate function.

The outward displays of morale are demonstrated through employees' respect and confidence in their leaders, involvement and desire for team unity, confidence in their fellow workers, and willingness to move forward even against tough barriers.

As mentioned in Rule 7, one of the increasingly used management tools to foster inspired collaboration and manage the informal interactions among divergent teams of employees is *social network analysis*, which maps the interacting networks of groups and individuals. Useful for small or large organizations, the tool has the following advantages:

1. Social network analysis exposes any conspicuous gaps where groups are not interacting—but should be.
2. Network maps point out where the frequent creative activities take place and offer concrete information to manage interactions.
3. The analysis shows the invisible, informal connections among people that are missing on a traditional organizational chart.
4. Social network analysis is excellent at providing valuable input about pinpointing individuals who consult with others most often, whom they turn to for types of expertise, and who either boosted or drained their energy levels.

With the current emphasis on maintaining a momentum of innovation and sustaining morale, if you map the informal networks you are better able to manage talent and nurture your organization's most expert employees. Such companies as **IBM**, **Capital Financial Corp.**, **Procter & Gamble**, and **Merck** use social network analysis to spur innovation, spot talent, and share information among diverse groups that do not normally work together.

In sum: Superior organizations and effective leadership produce powerful results—not only in thinking up new and better products but also in creative ways by reinventing themselves and maintaining the needed momentum just to survive.

Barriers to Building Morale

Undoubtedly and realistically, barriers to building morale will appear. The most significant ones are discussed next.

> *Lack of planning skills.* Planning within a team setting is a unifying activity. Therefore, it is in your best interest to hone those skills that foster a collaborative approach. There are numerous formats for developing plans. Most have similar characteristics. Following any of the popular formats or a standardized one that already exists in your firm should suffice. (See a format for a strategic business plan in the appendix.) Yet, within the context of building morale and unity, just acquiring planning skills is not enough. Rather, the more relevant issue is the team's ability to think strategically. That means

honing skills that permit you and others to look forward at least three to five years and then be prepared to act tactically by initiating actions for the short term. Further, thinking strategically translates to the effective application of strategy, which is the central purpose of this book. Strengthening planning skills is the device that opens your mind and those within your team to abundant possibilities.

Lack of drive. If your team is too easily discouraged, you can use the action steps that follow to activate morale. In some instances the morale problem radiates from poor leadership skills, which calls for a self-examination to identify where the weaknesses are embedded. In other cases, remedying the faults may be out of your hands, due possibly to the lack of involvement from the highest levels of management, compensation problems, career advancement concerns, and similar issues.

Lack of resourcefulness. This translates into being short of imagination, innovation, and all those factors that are generally understood to mean creativity. Herein is the crux of any firm's problems and consequently its opportunities—namely, to create innovative products, outfight and outperform competitors, and win firm footings in growth markets.

Lack of vision. This barrier to morale can be addressed for the most part by getting individuals actively involved in developing strategic business plans. Here, again, standard formats exist. Most of them provide for developing a mission statement or strategic direction, which is the same as creating a vision for the firm, business unit, or product line. The companies that have become leaders in their respective fields have mastered the process of developing visions of who they are and who they want to be over the long term. The most noteworthy ones include **General Electric**, **Procter & Gamble**, **Google**, **Home Depot**, and **Wal-Mart**.

Lack of self-development. One of the identifiable characteristics of a well-run organization is where a workable system exists for ongoing learning, which leads to career advancement. Self-development is certainly part of the morale factor and includes all the ingredients that go into a company's value system. This category aligns with the intrinsic importance of its human capital.

Lack of self-confidence. This morale barrier shows up in a manager's indecisiveness, lack of vision, overcaution, and unwillingness to attack a problem or pursue an opportunity with vigor and boldness. There are too many visible displays of such negative behavior to hide this human flaw from employees. In turn, like an epidemic, such damaging traits can spread to all those who are exposed to them. If the lack of self-confidence cannot be converted to a positive attitude through self-motivation and outside assistance, the manager should step down before any further damage is done.

Lack of communications skills. Poor communications ranks high as a major deterrent to maintaining morale. You should find out which forms of communication maximize interactions. The choices are many, from electronic devices to one-on-one contact. For instance, to encourage productive communications, **BMW** relocates many of its engineers, designers, and managers to its central research and innovation center to design cars. Face-to-face team contact reduces late-stage conflicts and speeds development times.

Procrastination. This barrier is the enemy of speed and is the sure way to dampen morale (see Rule 3). Quickly changing consumer demands, global outsourcing, open-source software, near instant movement of capital across oceans, electronic transfers of knowledge, and rapid communications from the field to decision-making executives at headquarters make procrastination an obsolete, and even dangerous, managerial trait for anyone responsible for committing resources and implementing strategies.

Erratic behavior. Of grave concern is volatile behavior that is inconsistent with a market situation. Employees can understand the need for a flexible managerial style or even a certain amount of eccentricity, if it is understood and accepted as part of a manager's inherent personality. However, they are unable to tolerate inconsistency and sudden erratic mannerisms, particularly if there is no apparent reason for what could be perceived as chaotic behavior.

Action Steps to Build Morale

In contrast to the preceding barriers, the following list provides concrete approaches to build morale:

Communicate. A foundation requirement to activate morale is communicating effectively up, down, and across levels. Doing so puts problems and opportunities out in the open and spurs the free flow of ideas to solve problems, attack opportunities, encourage others to think creatively, and surface any detrimental issues that would hamper a team's progress.

Encourage questions. Welcome them and use them. If you avoid questions, you create a damaging vacuum by preventing new possibilities from surfacing from those who would be inspired to act. Questions often trigger the creative process and open minds to think in areas not otherwise considered.

Delegate. Morale will rise if you challenge teams and give individuals responsibility. Often they will surprise you with solutions to problems that may have eluded you. Delegating also sends a strong message of confidence that you can let go and permit the team's creativity and initiative to take over.

Anticipate. Create an environment of expectation and excitement, such as the physical surroundings described in the **Johnson & Johnson** case. It should also be your intention to set up the psychological environment that encourages, and equally importantly, anticipates success.

Prioritize. With input of others, prioritize and rank those opportunities worth exploiting and problems needing solutions. Not only do you maximize resources, time, and energy, but you also involve the team in viewing the big picture and understanding the reasons behind the prioritizing.

Set high standards. Benchmark your activities and goals against the "best." Look beyond your company. Look to other industries for ideas and inspiration and show the team what can be accomplished. Try to draw meaningful parallels and let the competitive spirit flourish.

Show genuine interest. Be available. Listen to employees' problems, complaints, and other issues. Fundamental to building morale is demonstrating to your people that you are genuinely interested in them and their growth. This means being sincere and treating them as valued employees who have insight, experience, and special knowledge.

Respect differences. Accept the viewpoint that others may not see things as you do. Therefore, recognize differences of opinion, especially where there are diverse backgrounds. This is particularly so where more and more interactions take place with individuals in offshore locations.

Be candid. Within the range of what is permissible or confidential, explain management's actions to your employees. Provide information and guidance on matters that affect their security. (Review Herzberg's concepts cited earlier.) Keep your people up to date on all business matters affecting them and tackle unfounded rumors before they do damage.

Stay positive. Avoid displaying anxiety or worry. During times of intense pressure, sidestep disclosing your concerns or fears to your personnel. Instead, present a problem as a challenge that can be turned to an opportunity through a unified team effort.

Monitor performance. The object here is to stifle problems and correct situations before they fester. This process is also the basis for further training and an agenda for group meetings.

Morale Interfaces with Innovation

Innovation is the new differentiator. Therefore, use all your efforts to keep jobs stimulating. When something significant results from employees' creativity, express appreciation publicly for jobs well done. Your approach is to be evenhanded, open minded, and fair minded. Further, in all initiatives associated with innovation, support your people, communicate a vision of growth, provide them with the resources to do their jobs, and motivate them to learn and grow. It is as important to stand by them if something goes awry. Otherwise, you shut them down—along with creativity.

Also work at cultivating your group's miniculture. **3M** safeguards its culture by enlisting old-time employees to recount company history and hand down stories of 3M's long traditions for innovation to new engineers. Expanding the technique throughout the organization, after a while, every new employee is able to recite the foundation precepts that are the underpinnings of the organization.

Of course, be sure that what you focus on with your group complements the overall organizational culture. In doing so, you will find that parts of the culture deal with the work environment and issues related to quality of life—that is, do not expect innovation and performance from troubled and discouraged individuals with "life" issues. What follows is for you to establish criteria for evaluating innovation. You need to let the staff know how they will be measured. The frequently used measures for innovation include:

- Overall revenue growth attributed to innovations
- Percentage of total sales from new products or services
- Level of customer satisfaction
- Improvements in market share
- Comparisons of innovations measured against chief competitors by product category and market segment
- Actual number of new products or services launched
- Ratio of new product successes
- Other criteria required by your company

Quick Tips: Why Morale Suffers

- Employees are often skeptical of management's true commitment to meaningful change.
- The work environment does not encourage innovative thinking.
- Employees have little or no confidence that their opinions and ideas will be given any meaningful attention, with the likelihood that they will be dismissed without explanation.
- Employees lack trust in the leadership.
- De-motivators exist, such as job boredom, unfairness, casting blame, barriers to promotion, lack of recognition, and clamping down on mistakes.
- Executives fail to view self-esteem as a powerful motive to move individuals forward. Individuals do not wish to pass for cowards in the eyes of their peers (see Maslow's hierarchy list).
- Little effort is given to the dynamic influence of unity, which is established by discipline (as already pointed out in Rule 7) and by a supporting organization.

Morale Links with Trust

Decisive action builds trust and supports morale. So, too, are the manager's supportive words of encouragement and the projected image of reliability, integrity, and competence. These expressions should be activated by well-intentioned motives that are mutually serving, rather than self-serving.

Festering below trust, however, is the sometimes explosive behavior of subordinates, which is due to the impulsive nature of people. They nervously shrink back and find danger in any effort in which they do not foresee any possibility of success. Personnel also tend not to act as passive obedient instruments, but rather as very anxious and restless individuals who wish to finish things quickly and know in advance where they are going.

The purpose of focusing on trust, therefore, is to overcome these negative feelings among individuals who, for a variety of life experiences, mistrust and doubt. Some have valid reasons due to having been misinformed and misled in the past. Nonetheless, it is your challenge to create trust and convince them that you have their best interests at heart.

Summary

If you want to manage outstanding individuals, do everything to excite their ambition, get them to feel the excitement of accomplishment, and feel the dynamics of teamwork. Make them see a future with which they can identify because, even where there may be some unpleasant assignments, they will elect to do the job. It is within these jobs that you will find high-performing employees. In the last analysis, success is a matter of morale and your ability to engage heart, mind, and spirit.

The following Rule 9 emphasizes strengthening your decision-making capability. Before moving on, however, review Rule 8 using the strategy diagnostic tool to assess how this rule would affect your strategy.

* * *

Strategy Diagnostic Tool

Strategy Rule 8: Create a Morale Advantage

Part 1: Indications That Strategy Rule 8 Functions Effectively (Contributes to Implementing a Successful Competitive Strategy)

1. Managers actively seek to heighten morale and use it to energize the staff, tap their inner strengths, and get them to act decisively under competitive pressures.

    ☐ Frequently   ☐ Occasionally   ☐ Rarely

Create a Morale Advantage ▪ 167

2. Managers recognize that high morale holds a team together, affects day-to-day employee performance, and contributes ultimately to how well a business plan is implemented.
   ☐ Frequently   ☐ Occasionally   ☐ Rarely

3. Managers understand that even with cutting-edge technology and outstanding business strategies, their efforts will languish if employees are not roused with the driving will to succeed.
   ☐ Frequently   ☐ Occasionally   ☐ Rarely

4. Successful managers do the following to boost morale: emphasize ethical standards of behavior, encourage constructive communications, reward individuals for meaningful performance, and use a cross-functional team to encourage innovation.
   ☐ Frequently   ☐ Occasionally   ☐ Rarely

5. With morale affecting day-to-day employee performance, managers try to create a work environment that fosters creative thinking and encourages employees to offer opinions and ideas—with the assurance that they will be given serious attention.
   ☐ Frequently   ☐ Occasionally   ☐ Rarely

6. Managers move rapidly to overturn any skepticism shown by employees about the organization's commitment to meaningful change, and they do so before it festers into a morale problem.
   ☐ Frequently   ☐ Occasionally   ☐ Rarely

Part 2: Symptoms That Strategy Rule 8 Is Functioning Ineffectively (Detrimental to Implementing a Successful Competitive Strategy)

1. Employees visibly display a lack of respect toward their leadership.
   ☐ Frequently   ☐ Occasionally   ☐ Rarely

2. Employees feel insecure about their jobs and have doubts about management's concern for their well-being.
   ☐ Frequently   ☐ Occasionally   ☐ Rarely

3. Some key managers do not seem to be concerned about morale and employees are generally left to fend for themselves.
   ☐ Frequently   ☐ Occasionally   ☐ Rarely

4. There is no sign of a cohesive team spirit.
   ☐ Frequently   ☐ Occasionally   ☐ Rarely

5. There is a conspicuous lack of new ideas, suggestions, or opportunities bubbling up to management, as well as a noticeable absence of a workable two-way communications system.
   ☐ Frequently   ☐ Occasionally   ☐ Rarely

6. No consistent procedure is present to encourage and reward employees for innovative suggestions.
   ☐ Frequently   ☐ Occasionally   ☐ Rarely

7. Employees exhibit erratic behavior, such as discouragement, indifference, fear, and resistance to taking all forms of risk.
   ☐ Frequently   ☐ Occasionally   ☐ Rarely

8. When new market initiatives and competitive encounters have generally failed, management routinely lays blame primarily on poor employee performance.
   ☐ Frequently   ☐ Occasionally   ☐ Rarely

The ratings for Parts 1 and 2 are qualitative assessments of managers' overall ability to use morale to execute an effective competitive strategy. Based on a diagnosis of your company's situation, use the following remedies to implement corrective action:

- Communicate a long-term positive vision for the organization with clearly stated objectives that employees can internalize and get excited about.
- Develop a learning environment that demonstrates to employees that management is interested in and supportive of their development.
- Remove any personal and physical barriers that would prevent employees from gaining pride in their work and the organization.
- Create collaborative cross-functional teams with specific duties and responsibilities.
- Encourage constructive communication up and down the organization where self-expression and innovation are encouraged.
- Nurture a corporate culture that is reinforced by an ethical climate and secured by the company's positive history and core values.
- Utilize such tools as social network analysis to identify where interactions take place and with whom.
- Require teams to submit strategic business plans, which serve as a unifying activity to allow collaboration and team dynamics to flourish through creative expression.

Remedies and actions: _____

Finally, to demonstrate the far ranging impact of this rule, the following company problems linked to the use of morale come from a survey of chief executives of medium and large organizations (company names withheld to maintain confidentiality):

"With consolidation throughout the industry and a more competitive environment, there is a need to squeeze savings, cut jobs, and close branches. Yet, it's imperative that we keep employee morale up and reassure customers that they won't be abandoned."

"We've stumbled badly because of shoddy quality. Now we must get our people behind the product line, play down the tarnished name, improve quality, and set prices a bit below the market leaders."

Strategy Rule 9

Strengthen Your Decision-Making Capabilities: Fortify Intuition, Enhance Business Experience, Expand Knowledge

Chapter Objectives

1. Interpret business situations despite the fog of uncertainty.
2. Rely on intuition to reach reliable decisions.
3. Use case histories to enhance and broaden your business experience, knowledge, and training.
4. Employ the strategy diagnostic tool to assess Rule 9.

Introduction

From the onset of most business campaigns, the unfolding events related to entering a market, introducing a new product, or confronting an aggressive competitor are fluid and cloaked in uncertainty. Even your follow-up actions can be preplanned only to a certain point with any reasonable chance of accuracy. The only constancy resides within the depths of your natural intuition, accumulated experience, acquired training, and amassed knowledge.

Your remarkable and potentially powerful inner self is your only trustworthy channel for penetrating the fog of market vagueness. It is the means by which you can reliably influence events with the solid conviction that your decisions have a reasonable chance of being correct. Fortunately, these internal forces usually surface at the moment when you most need direct action.

Exactly what market uncertainties should you take into account? Consider the haziness that surrounds the following questions:

What direct or indirect strategies will competitors likely use to challenge your market-entry efforts? What perceptions will customers bring into play about your new product—positive, passive, or negative? Will they accept or reject your promotional themes? What are the chances of offshore competitors preempting your offerings with low-price knockoffs?

What environmental, legal, and political forces might interfere with your moves? How dependable is your supply chain in handling transactions and the physical movement of your product? Is there sufficient after-sales service in place to satisfy customers? Is the sales force adequately trained and motivated to handle the product introduction?

These varied and tough-to-answer questions force you to activate your innate systems. It is in this murkiness of unknowing and not being able to accurately assess your current situation where you must rely on spontaneous decisions. These are the conditions in which intuition, experience, and training flourish.

In the end, it is the mind that wins competitive encounters, has always won them throughout all business history, and will continue to win them in every campaign where the human element is present. Accepting this truism should lead you to three concurrent approaches:

1. Rely heavily on your experience and steady stream of information about the marketplace.
2. Broaden your business knowledge, with particular emphasis on studying actual company examples, current and past, and extracting from such study any promising strategies that could possibly apply to your situation (see Exhibit 9.1).
3. Rely on your intuition as a trustworthy source for answers to guide your actions.

The likelihood, however, is that you depend primarily on a more singular pathway of personally acquired knowledge, which may be long by your own standards. However, within a broader framework, your working life is somewhat short and your experience is shorter still at a particular job level or function.

Exhibit 9.1 Sampling of Actual Company Problems[a]

> How do we deal with offshore competitors selling into our market with prices 30 percent to 45 percent below ours? (Cummins Engines)
>
> With large multinational organizations tending to dominate our market, what strategies are possible? (Xerox)
>
> Commodity products—and those in mature markets—often end up in severe price wars. What strategies can we use to differentiate our product and reduce pressure on price? (Georgia Pacific)
>
> How do we play catch-up after lagging behind aggressive competitors for years? (Texas Instruments)
>
> How do we position our products against established market leaders without igniting hostile competition? (Tandy Corp)
>
> Corporate culture is the new buzzword. But how do we apply corporate culture to achieve competitive advantage? (Mitsubishi Electric)
>
> What strategies are practical when the market leader repeatedly retaliates against every action we undertake? (Fuji Photo)
>
> How do we implement a niche marketing strategy? (Williams-Sonoma)
>
> What can we do when competitors and customers attempt to reposition our distinctive product to a commodity status? (Whirlpool)
>
> How do we develop a long-term vision for our product when day-to-day problems seem so overwhelming? (Air Products & Chemicals)
>
> How do we rebuild our brand's image and win back customers who were once behind our product? (AT&T)
>
> How do we revitalize a product line and keep it from becoming an also-ran in the industry? (Timex)
>
> How can we find a point of entry in a market dominated by industry leaders? (Tecnol Medical Products)
>
> How do we cope with the possibility of our product becoming a dinosaur? (Moore Corp.)
>
> How do we cope with erratic marketing and loss of sales momentum? (PPG Industries)
>
> How do we recover after an aggressive competitor surprises us by launching a new product to our customers? (Nabisco)
>
> What defensive strategies are effective to protect our market share? (Haworth Furniture)

[a] The problems and company references all come from generally available business publications. It is therefore appropriate and highly useful to assign an individual the task of scouring the business press for examples to build a business history database.

Even when you move ahead to new responsibilities, those who follow you may not have access to your knowledge, just as you may not have benefited from the practices of those you followed. Yet, to gain any value from the hard-earned experiences of you and others, it is valuable to capture and retain lessons that are transferable to your company or function. These include details of your company's current situation, the history of your firm and the industry in which it operates, and any insights derived from other fields. Otherwise, the accumulated knowledge is lost and has to be recreated by the next person occupying the position.

For most experienced managers, the lessons of yesterday exist solely in their memories. Nonetheless, they should be able to provide the valued know-how to others. In Rule 8: Create a Morale Advantage, you saw how new employees at **3M** were systematically exposed to stories about the company's impressive tradition for innovation by long-term employees, who also told of the company's history, morals, and deeply rooted beliefs. The only caution here is to make certain that the information passed along is not overly opinionated with commentary that may stifle innovation with counterproductive comments such as, "We've tried that before and it didn't work."

For most companies, however, experienced individuals rarely have opportunities to speak, write, or instruct others. Thus, valuable knowledge is often lost when they leave the organization or get promoted to levels where face-to-face contact is missing. The essential point is that, looked at through the eyes of an experienced manager, even the smallest detail about an actual marketplace incident passed along to another person can be extremely instructive.

Former CEO Jack Welsh's frequent appearances and lectures at **General Electric**'s internal training sessions offered valuable ongoing streams of insightful information to midlevel managers. The training format offered a convenient setting to penetrate numerous layers of management and circumvent the years of experience between senior management and lower level personnel who were entrenched with day-to-day front line problems.

Consequently, initiate (or recommend) procedures to capture the insights, knowledge, and observations of numerous individuals and categorize them into usable databases that would be available and easily accessed through training, mentoring, written documentation, or oral exchange.* Even where you cannot affect change in the organization, it is possible to take the initiative and set up a system within a small group (see discussion ahead on managing knowledge).

These points are illustrated in the following case example.

Case Example

Nucor Corp. produces high-grade steel and does it efficiently and profitably against its rivals. Even benchmarked against almost all other

---

* In no way should this process be confused with a typical transition memo, which tends to be more limited in scope and deals with fundamental administrative procedures and the like.

companies in the Standard & Poor's 500-stock index, Nucor's 387 percent return to shareholders over the past five years handily beats most other companies, including New Economy icons **Amazon.com**, **Starbucks**, and **eBay**.

Nucor has become more profitable as it has grown: Margins, which were seven percent in 2000, reached ten percent in 2005. Between 2000 and 2006, sales skyrocketed from $4.6 billion to $14.8 billion. What is behind such exemplary performance in an industry as Rust Belt as they come? The answers fall into several categories:

1. Performance

    On average, in a nonunion environment, two thirds of Nucor steelworkers' pay is based on a product bonus. This measure of performance is anchored to management's understanding of human behavior. Employees, even hourly workers, will make extraordinary efforts if rewarded richly, treated with respect, and given real authority. At the executive level, compensation is also tied to product performance. As an add-on, executives are measured by tangible actions at beating competition and outpacing a sample group of other high-performing companies.

2. Corporate culture

    To assure a cultural compatibility throughout the organization, a managerial priority is to instill Nucor's unique culture in all of the 13 plants it acquired over a five-year period. Teams of highly experienced veteran workers visit with their counterparts in newly acquired facilities to explain the system face to face. Part of that cultural process includes explaining the deeply rooted custom of providing a helping hand without obtaining approvals from supervisors. It is a common everyday occurrence, for instance, for employees in one plant to take the initiative and help others in sister plants to get operations up and running should severe problems shut down production, regardless of time, distance, or any inconvenience. Even the cultural symbols are considered important. Every year, for example, all employees' names go on the cover of Nucor's annual report.

3. Entrepreneurship

    Nucor's flattened hierarchy and emphasis on pushing power to the front line leads employees to adopt the mindset of owner–operators. That focus gives free reign to encouraging employees' imaginations and intuitions to flourish. For example, the development of thin slab casting of sheet metal has made Nucor an industry leader. In another instance, employees in one plant had to innovate themselves out of a predicament. This plant's particular form of steel could not be produced profitably any longer. Relying on their personal knowledge, experience, and intuitive creativity, employees found types of specialized steel they could produce more profitably that would be less threatened by imports.

4. Motivation

There is unwavering attention heaped on employees. That means talking to them, listening to them, taking a risk on their ideas, and accepting an occasional failure. It is about nurturing their experiences and sharing them with others in a proactive way. Nucor systematically sends new workers to existing plants to hunt for improvements. Older workers also travel to newly acquired plants to find out what they can learn. These experienced individuals actively tune in to sharing ideas and experiences, as well as stay alert for innovations they can take back to their home plants. There is a healthy competition, too, among facilities. For instance, plant managers routinely set up contests to try to outdo other plants in areas such as safety, efficiency, product quality, or output. It all ties in with the company's long history of cooperation and idea sharing.

The Nucor case provides a learning platform for you to strengthen your personal decision-making capabilities. The following categories hone in more precisely on developing that capability: *valuing business history, managing knowledge,* and *activating intuition.*

Valuing Business History

What can you learn by taking a more expansive view of business history? Initially, you get to see your problem through a more discerning lens, whereby you are able to define your situation with greater precision. The clarity is provided by looking at actual case examples of businesses in and out of your industry. Through that medium you will find problems that correspond to your actual condition, or at least come remarkably close to your situation. You can then consider adapting those strategies to your own needs or use them to stimulate your thinking and originate new ones.*

Thus, by overlaying your situation on similar business circumstances, you may see commonalities that cause you to think more critically about your problems.

---

* The intent here is not to copy another's strategy. Rather, the purpose is to recognize that similar problems do exist in other companies and in other industries. If the strategy fits the symptoms of your problem, has proved successful, and parallels the strengths and resources of your organization, consider taking similar action.

Exhibit 9.1 lists problems with actual names of companies that confronted those situations. (Can you find any competitive situations that come close to what you currently face?)

As important, case histories offer a broader vantage point that allows you to see causal events that may have triggered your problem. In turn, that view would provide an entirely fresh perspective about what you may be facing. The process opens your mind to ask pertinent questions and to receive credible answers about what went right or wrong.

For instance, in a postexamination of an event, such as an unsuccessful entry into a new market, you can seek answers to the following questions:

- Did our failed product launch result from a faulty strategy at the onset, such as attempting a direct confrontation against a more powerful and entrenched competitor?
- Did we incorporate sound strategy principles into our business plan, such as speed, indirect approach, maneuver, concentration, alternative objectives, and unbalancing the competitor?*
- Did our strategy include a unique competitive advantage in product, service, technology, price, promotion, or in the supply chain?
- Did we actively make use of competitive intelligence from a reliable source or were our decisions based on hearsay and unsupported assumptions?
- Can we recount events and determine if the resources expended achieved our planned objectives or were resources used as a battering ram with little market penetration to show for the effort?
- Was the strategy cost effective?
- Did the commitment take valuable and perhaps scarce resources away from other avenues of growth? Were we left vulnerable in our core markets?
- How would we gauge the competency with which the plan was implemented? Was it vigorously and enthusiastically carried out with a spirited, energetic approach or did individuals lose morale and sag under competitive pressures?

It is in your best interest to expand on the last question and examine the cultural and psychological aspects of employee behavior, with special attention to morale (see Rule 8). The implication is that a dispirited and unmotivated effort will carry with it a tremendous psychological burden that could put your efforts at risk—regardless of the amount of money you pump into the effort. Furthermore, if an aggressive competitor challenges you with a more energetic push, any negative morale and psychological issues at your end could result in only minimal success and an inability to recover valuable resources.

Therefore, systematically examining other companies' problems, as listed in Exhibit 9.1, would provide you with substantial guidance. As you scan the problems,

---

* All of these strategy principles have been explained in previous chapters.

notice an overall commonality of conditions that fall into categories, such as competition, products, falling sales, maturing markets, and the like. (Can your problems be all that much different?) Thus, valuing business history by examining actual case examples, regardless of timeframe, permits you to develop an invaluable body of practical knowledge as a resource from which to build additional layers of knowledge, insight, and experience. In turn, you feed the intuitive process and increase the possibilities for reaching a successful solution.

As already noted, you are cautioned not to force-fit someone else's unique solution to your problem. Rather, the aim is to extract any universal principles that would apply or call to mind ones you may have overlooked. There is a strong likelihood that you can adapt, with modification, a strategy for your situation. You will find that assembling case histories broadens your entire base of knowledge, takes you out of the narrow lane, and adds a fresh perspective to your thinking that will surface as insightful decisions at the proper time.

Managing Knowledge

As a backdrop to fortifying your experience, intuition, and training, you can view knowledge management (KM) as an organized and more elevated form of information. When you interpret a business situation, organized knowledge provides new meaning, value, and relevance to your decision making. Knowledge management consists of two parts:

- *Explicit knowledge* exists in your internal databases, records, manuals, documents, the raw numbers in spreadsheets, and increasingly through Web sites. In particular, sophisticated analytics incorporate additional data from a variety of customer touch points, such as call centers, field services, prior marketing programs, and external back-office information. When organized into a useable format, there is greater reliability and accuracy for estimating a market situation.
- *Tacit knowledge* generally resides in the minds of individuals who have accumulated it through discovery, experience, or intuition, or through numerous interactions with others. Since tacit knowledge tends to be less structured, it cannot always be put down on paper. Instead, it is transferred indirectly through conversation, observation, or other types of informal interchange. Tacit knowledge can originate in a variety of patterns, such as the impressions, feelings, and insights of a sales rep returning from a visit with a key customer. It can start with an engineer making an offhand comment about a gestating idea with an associate in a casual setting over lunch.

If given the same level of seriousness and discipline as with any business system, KM can operate as a balanced, multidisciplinary framework for capturing, sharing, and spewing forth immensely valuable knowledge. Consequently, in your key managerial role of making a diverse range of decisions, actively involve yourself in developing an internal KM exchange network. Regardless of the size of your company, it should blend with the everyday life of your organization and feed the transfer of meaningful knowledge, including the use of case histories discussed before.

For many organizations explicit knowledge is tangible and available on a widespread basis or, minimally, it is accessible to several layers of personnel. On the other hand, tacit knowledge is somewhat unbounded and tends to be used by individuals who need to protect what they know as a personal defense or a power barrier. As indicated earlier, it is this form of knowledge that is often lost to others, if it is not captured, organized, and accessible.

To break through the barriers and add KM to the culture of the firm means establishing a level of trust up and down the organization and instilling a spirit of teamwork to make knowledge management work to the full benefit of the organization. By blending explicit and tacit knowledge, more accurate decisions can evolve. For instance, you would be in a far more advantageous position to justify (or recommend to a management committee) the expenditure for developing, and rolling out a new product, adopting a cutting-edge technology, or probing an evolving market segment. Most importantly and in the context of this book, you can shape strategies and tactics that sustain a competitive edge.

The responsibility for KM can reside with any number of individuals, depending on the size of your company and the level of priority given to the project. The range of individuals includes company librarian, information technology (IT) manager, market research manager, or the new title of "internal infomediary," who creates or manages systems to connect employees with the knowledge they need. Regardless of title, the central responsibility is to tap into the immense fund of knowledge flowing around the organization.

Finally, information, intelligence, and total knowledge management are only as valuable as your willingness to apply the mass of knowledge to make appropriate decisions and create a winning business advantage.

Activating Intuition

Whenever pushed to make a decision with less information than you would like, which is usually quite often, you are likely to rely on intuition, or gut feeling. Whether or not you are consciously aware of it happening, there are few options open to you, unless you depend on others to decide for you.

Think of intuition as an inner voice that presents you with possible courses of action. It unlocks the mind and guides you by means of a flash of insight, ideas, images, metaphors, or symbols. "The intuitive mind tells the rational thinking mind where to look next," declared the renowned Dr. Jonas Salk.

While intuition is often thought of as something elusive, spontaneous, and outside our control, it is nevertheless possible to make intuition more accessible and more reliable. It begins by paying attention to your feelings. This means starting with a quiet mind and turning off the constant monologue that clutters the mind. You then adopt a mindset where simply *knowing* transcends reason. Just perceiving possibilities is also an intuitive function. Of course, you can never be certain what the outcome of a decision is going to be. But you can have a strong intuitive sense of the direction you want to pursue.

Yet such a leap of understanding is not in opposition to or a substitute for reason, and it certainly is not in conflict with the input of reliable market intelligence. Intuition is inner power that you use in addition to reason and factual information. In effect, there is an integration of intuition with the logical and linear thinking mind.

To make intuition work, you have to learn to hold attention on something for more than a few seconds, which can be a real challenge internally. For some, that also means emptying the mind of the emotional baggage that upsets, angers, and creates a chaotic state of mind. It means letting go and being nonjudgmental. You may have to personalize how you reach the receptive state where intuition can flourish. It may be through meditation, solitude, being out in nature alone and quiet, or using slow breathing techniques.

The object is to quiet the feelings, quiet the body, quiet the mind, and be left with a kind of inner knowing. In the silence you learn the most about intuition. It is in this period of incubation that you let a problem just sit and you take time out. Other approaches include using mental imagery, which is more associated with thinking and tends to be the type of intuition used by business executives, especially entrepreneurs, who tend to be highly intuitive.

In all methods, a common requirement is to trust yourself. Then you can trust your intuition. Keeping track of the accuracy of your intuitive decisions will give you an indication of whether your mind is prepped for intuition to thrive.

The benefits come when you can perceive possibilities in the future. In its most practical business application, intuition applies to developing a strategic business plan where the first planning step is to develop a mission statement for your company or business unit or to define a long-term strategic direction for a product line.

Anything that is creative, breaks new ground, provides a future-directed vision, or pushes you beyond the boundaries of what you already know is intuitive. If any mystery exists about intuition, it is that people seem to get information and they do not know how they got it. Mathematicians, for example, can arrive at theorems that have never been proven before, just through their intuition. "No problem is solved by the same consciousness that created it," stated Albert Einstein, who was well known for his use of intuition.

What is required is a shift in consciousness; through this shift, you tap the subconscious mind to show you the way to solve a problem, which is revealed as intuition. What follows is to use logic, reason, and market research to follow up on the intuition for proof and validation. However, if time does not permit that luxury, then you must react with trust in your intuition.

The creative leap is always an intuitive one that enables you to see things that you have not noticed before. It is a new perception, as though intuitively you notice what you have not noticed before or you acknowledge what you already know but have forgotten. It is a sense of inner knowing rather than something that you need to learn about. Again, being intuitive is to trust in yourself. We are all equipped with intuition. It is built in. It is pragmatic. It solves problems. It identifies opportunities that may not be seen with conscious vision.

The following case illustrates the preceding concepts.

Case Example

Whirlpool Corp. faced a series of severe problems, notwithstanding its lofty position as the world's No. 1 maker of big-ticket appliances. The central problems were that Whirlpool's machines had been reduced to commodities and that prices for its most important products were falling each year. As one Whirlpool executive explained: "I go into an appliance store. I stand 40 feet away from a line of washers, and I can't pick out ours. They all look alike. They all have decent quality. They all have the same price point. It's a sea of white."

But that explanation described only a surface issue. Underlying the situation was Whirlpool's history of giving relatively minor attention to new product development and innovation. Whirlpool was good at operating plants and distribution channels efficiently, and it was outstanding at turning out quality washers and dryers that were solid and long lasting. The company also achieved unmatched economies of scale with its perpetual cost cutting. Only on an as-need basis did engineering tweak its products to lower costs or boost performance, such as by better insulating a freezer or adding another washing cycle. But that was about as far reaching as product development got.

Challenged by the flagging profits and flat sales, Whirlpool made a series of turn-around moves. Its strategies focused in two areas. First, it needed to develop strategies that would prevent competitive imitators from moving its product line into a commodity status, where price wars become the central marketing focus. Second, the company needed to reinvent the corporate culture so that innovation and creativity became the essential strategies as well as the trigger to energize the company plagued by missed opportunities, sluggish sales, and dismal profits.

But how is the changeover accomplished? How do you teach people to be creative? Whirlpool moved forward with the following actions:

An intranet site offered personnel a do-it-yourself course in innovation and listed every project in the pipeline. Employees were then invited to post ideas or to network informally with others to gain their expertise.

The company hosted innovation fairs to salute inventors and elicit more ideas. Senior executives continued to encourage workers to go to their bosses with proposals. Whirlpool's "knowledge management" site recorded up to 300,000 hits per month.

Guidelines were provided about how to present ideas that would improve the chances of surviving an executive review for funding. New ideas would have to enhance the company's existing brands or products. Only then would approved projects go to representatives from design, market research, R&D, and manufacturing departments for implementation.

Projects would be analyzed and validated through market research to determine if the new product would likely command above-average markup.

Products that failed the first go-round are viewed again for revival. The intent is not to kill ideas. Instead, these projects are shelved for a period of time so that other employees can take a look at them later in a more insightful mode of thinking.

Some of the products that have emerged include an innovative, feature-loaded waffle iron that is three times costlier than the old waffle iron; an all-in-one gas grill, refrigerator, oven, stereo, and bar that fits onto the back of an SUV for tailgate parties; and a duet washer and dryer set with a variety of innovative accessories. In five years, new products that fit Wirlpool's definition of innovation have skyrocketed from $10 million to $760 million.

Beyond the success in achieving a resounding turnaround in its fortunes, the grandest form of complements came from such companies as **Hewlett-Packard**, **Nokia**, and **Procter & Gamble**. These organizations, respected leaders in their own fields, eagerly sent groups to benchmark their own innovation programs against Whirlpool's.

What can you learn from the Whirlpool case? First, the merging of business history, experience, training, corporate culture, and intuition drives creativity and innovation. Second, as market conditions change, technologies advance, and competition hardens, many of the successful strategies from other times, places, and industries should be studiously reviewed for their potential value. Third, the challenge is for you to discover from each case example what useable information you can take away that would expand your knowledge and ability to analyze your company's situation. Specifically, you want to use the following process:

- State the central problem, as shown in Exhibit 9.1.
- Describe the details about the company's situation that you are examining.
- Indicate what actions the company took to solve its problem.
- Point out the lessons you can learn in the form of action strategies.

Fourth, recognize that the road you must travel to carry out your business plan can never be determined beforehand with absolute accuracy. Therefore, you have to be actively engaged, visible on the scene to observe events with your own eyes, and able to assimilate all the information. Then, you can follow through with an action plan. What you see must be accurately interpreted using your innate powers of knowledge, experience, training, and intuition.

Therefore, try to obtain all the details of the events. Do so by conducting a series of one-on-one interviews with individuals directly connected to the campaign. Also hold group debriefing sessions to acquire as much information as possible. This effort is likely to show in striking, irrefutable facts what was actually taking place at various times and places during the campaign.

Going through such details, along with your personal estimates of the individuals involved, serves as a foundation for what could lead to new strategies. In addition, that review should yield valuable insights related to changes in training or new methods of performing in the marketplace under a variety of conditions. Here is where a reliable business plan is essential to capture and organize all the diverse information.

Quick Tips: How to Combine Intuition, Experience, Training, and Knowledge into an Awesome Force

- Examine your past campaigns, events, and strategies, including those of competitors. Do not overlook the smallest detail that could help you figure out what went right or wrong in a campaign. Pinpoint those lessons that would be the basis for shaping future strategies.
- Tap the experiences of individuals who can recapture events of long ago, especially those who have been elevated in the organization, or, if possible, those who have left the company.
- Tune in to actual business case examples, past and present, in and out of your industry, for ways to enhance your skill and broaden your range of knowledge.

- Anticipate that your experience, knowledge, training, and intuition will appear as an idea, image, or a flash of insight. This can occur even under unfortunate circumstances when you may have given up hope of their influence.

The Human Element

If your employees get entangled in a tough competitive encounter, no doubt many will face stumbling blocks, which they will promptly pronounce as overwhelming. They may even complain that the possible solutions are too involved or overly fatiguing. Such reactions are understandable because, for some, there is the natural timidity of humans to see only one side of anything and thereby develop a first impression that usually inclines toward fear and exaggerated caution. (See the section on motivational behavior in Rule 8.)

If, as a manager, you give in to those complaints and frailties, you will soon succumb completely. Instead of acting with courage and determination, your efforts will be reduced to weakness and inactivity. To resist all this negativity, keep faith in your inner self. Focus on all your knowledge, training, experience, and, yes, intuition. At times, even with all the uncertainties of the marketplace, this mindset may appear as stubbornness. Actually, it is an expression of strength of mind and character, called firmness.

The reality is that there will be the inevitable miscalculations and hurdles that you are certain to face. Notwithstanding, if you pursue your aims with boldness and confidence, you will reach your goals in spite of obstacles. Where the law of probability is often the only valid guide, you must stand fast and trust in your honest interpretation about what is realistically possible and what is not. This is a managerial role that is not easy to play, yet one that is mandatory to successful performance.

Equally important is the trust you must have in your subordinates. Therefore, choose individuals on whom you can rely. Such reliance and confidence is directly proportional to the training they receive and the manner in which they are led. The following example illustrates these points.

Case Example

Carlos Ghosn has the mind-boggling job of serving as chief executive of both **Nissan** of Japan and **Renault** of France. He prods his employees at all levels by creating an intense feeling of urgency. If complaints crop up, he dismisses them and any other perceived difficulties with an irritated wave-off.

Strengthen Your Decision-Making Capabilities ▪ 185

The urgency is intentionally created as part of Ghosn's style of anticipating problems, putting them on the table, and dealing with them before they happen. His study of corporate history shows that if you wait too long to tackle problems, you are likely to face a tragedy. "Ghosn is absolutely tenacious in fighting complacency and the notion that we are in good shape," declares one executive. Thus, Ghosn functions as if collapse lurks around the next corner. He is fueled by a sense of crisis, mixed with impatience and passion, as he runs, simultaneously, two of the major car companies in the world.

Operating in a cultural environment where success breeds arrogance, Ghosn's approach is to elevate the creativity and productivity of workers and thereby manage the human element. Ghosn's leadership style provides a resounding justification for strengthening your resolve against the weakening impressions of the moment. Even fortified with personal decisiveness and armed with sound intelligence, you can still succumb to wrong decisions dictated by fear. You must not allow these errors to shake your faith or tempt you to accommodate to those fleeting impressions. The inevitable difficulties, therefore, demand your confidence, firmness, and conviction; whereas "ordinary" managers may find in obstacles ample excuses to give in.

The essential point is not to give ground in competitive encounters until the very last moment, after all options have been considered. If your plan is based on using sound strategy principles—such as boldness, indirect approach, concentration, and speed—and if you are resolute and persistent in implementing your plans and determined to accomplish your goals, then you will find success in your efforts. The steadfast lessons of this rule are:

1. Count on strengthening your decision-making capabilities by fortifying intuition, enhancing your business experience, and expanding your knowledge.
2. Look for the infallible connection of corporate culture and competitive strategy. It is the determining factor if you are to be successful in implementing your plan, directing your competitive strategy, and ultimately outperforming your rivals.
3. It is in the marketplace that an organization justifies its existence. Such justification, however, must be based on honesty and integrity to the foundation principles, value systems, and overall culture of your organization—as well as to the long-term development of the marketplace—and not rigged to any personal agenda or avoidance of factual market intelligence.

In the end, it is mastering the nine rules of strategy that can direct marketplace events in your favor.

Now review Rule 9 using the strategy diagnostic tool to assess how this rule would affect your strategy.

* * *

Strategy Diagnostic Tool

Strategy Rule 9: Strengthen Your Decision-Making Capabilities

Part 1: Indications That Strategy Rule 9 Functions Effectively (Contributes to Implementing a Successful Competitive Strategy)

1. Managers actively strengthen their decision-making capabilities by systematically examining the best practices of other companies and using them as resources to build additional layers of knowledge.

    ☐ Frequently ☐ Occasionally ☐ Rarely

2. When reliable competitive intelligence is missing, managers tend to rely with confidence on their innate intuition, experience, training, and knowledge to arrive at valid decisions.

    ☐ Frequently ☐ Occasionally ☐ Rarely

3. To enhance the decision-making skills of front-line employees, managers give them free reign to develop their imaginations and intuitions.

    ☐ Frequently ☐ Occasionally ☐ Rarely

4. Managers understand that by valuing their organization's business history and managing knowledge, they add greater precision to their decision-making skills.

    ☐ Frequently ☐ Occasionally ☐ Rarely

Part 2: Symptoms That Strategy Rule 9 Is Functioning Ineffectively (Detrimental to Implementing a Successful Competitive Strategy)

1. Management has no organized effort to capture and record the valuable lessons and key strategies of past events to pass them on to the next generation of decision-making employees.

    ☐ Frequently ☐ Occasionally ☐ Rarely

2. Key executives do not have a systematic procedure or venue to deliver live commentary to selected individuals about their experiences, insights, knowledge, and observations.

    ☐ Frequently ☐ Occasionally ☐ Rarely

3. Many staff members tend to flounder and become dispirited when faced with an intense competitive situation.

    ☐ Frequently ☐ Occasionally ☐ Rarely

Strengthen Your Decision-Making Capabilities ■ 187

The ratings for Parts 1 and 2 are qualitative assessments of managers' overall ability to enhance their decision-making capabilities to execute an effective competitive strategy. Based on a diagnosis of your company's situation, use the following remedies to implement corrective action:

- Provide training sessions where key executives are invited to provide their insights and experiences to attendees.
- Set up databases that capture case histories and details of significant business events, as well as noteworthy competitive encounters—both successful and failed ones.
- Permit individuals to access the assembled information for their personal use, to coach others, and for formal training.

Remedies and actions: _____

_____

Finally, to demonstrate the far-ranging impact of this rule, the following company problems linked to effective decision-making come from a survey of chief executives of medium and large organizations (company names withheld to maintain confidentiality):

"The business is floundering due to a lack of cohesiveness and focus."
"Merchandise isn't moving and customers are shopping elsewhere."
"We're zigzagging between two conflicting goals: market share or profits."
"We need experience in joint ventures, mergers, and other types of alliances if we're going to succeed as a global competitor."

Appendix

The Strategic Business Plan: Forms and Guidelines

Introduction

The strategic business plan (SBP) is the "housing" for all your objectives, market intelligence, and, ultimately, the competitive strategies that will set your plan in motion. With the forms and guidelines that follow, you can develop your personalized plan. You have the flexibility to customize the forms by inserting specific vocabulary and unique issues related to your industry and company, while retaining the planning structure of a proven format.

Also, you can make the SBP a permanent part of your management operating system by scanning this entire planning format into your computer. Doing so permits you to add special forms required by your organization or to insert any of the commercially available spreadsheet programs, as well as the growing number of customer relationship management (CRM) programs.

As you develop your SBP, keep in mind the rules of competitive strategy presented in this book. Use them as a continuing guide. They will help you strengthen the plan and improve your chances of endorsement from senior management; ultimately, they will increase the likelihood of achieving your planning objectives.

Overview of the Strategic Business Plan Strategic Section

You can obtain optimum results for your SBP by following the process diagrammed in Figure A.1. As you examine the flowchart, notice that the top row of boxes represents the *strategic* portion of the plan and covers a three- to five-year timeframe. The second row of boxes displays the *tactical* one-year plan. It merges the strategic plan and tactical plan into one unified SBP and makes it a complete format and an operational management tool to energize your company's potential.

You will find that following the SBP process will add an organized and disciplined approach to your thinking. Yet the process in no way confines your thinking or creativity. Instead, it expands your flexibility, extends your strategy vision, and elevates the creative process. In turn, the strategy vision results in providing you with a choice of revenue-building opportunities expressed through markets, products, and services.

Section 1: Strategic Direction

The first box in Figure A.1, section 1, *strategic direction,* allows you to visualize the long-term direction of your company, division, product, or service.

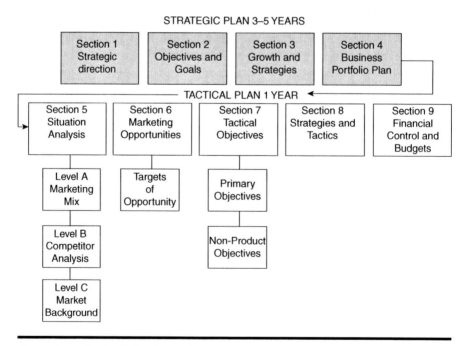

Figure A.1 Strategic business plan.

Planning Guidelines

The first step is to use the following questions to provide an organized approach to developing a strategic direction. Answering these questions will help you shape the ideal vision of what your company, business unit, or product or service will look like over the next three to five years. More precisely, it should echo your (or your team's) long-term outlook, as long as it conforms to overall corporate objectives and policies. To develop your strategic direction, answer the following six questions:

1. What are your firm's distinctive areas of expertise? This question refers to your organization's (or business unit's) competencies. You can answer by evaluating the following:
   - Relative competitive strengths of your product or service based on customer satisfaction, profitability, and market share.
   - Relationships within the supply chain or with end-use customers
   - Existing production capabilities
   - Technology advantages
   - Size of your sales force
   - Financial strength
   - R&D expenditures
   - Amount of customer or technical service provided
   - Capabilities of personnel
   - A corporate culture favorable to supporting the SBP

   Fill in: _____
   _____
   _____

2. What business should your firm be in over the next three to five years? How will it differ from what exists today?

   Fill in: _____
   _____
   _____

3. What segments or categories of customers will you serve?

   Fill in: _____
   _____
   _____

4. What additional functions are you likely to fulfill for customers as you see the market evolve?

   Fill in: _____
   _____
   _____

5. What new technologies will you require to satisfy future customer and market needs?

   Fill in: _____

   _____

   _____

6. What changes are taking place in markets, consumer behavior, competition, environmental issues, corporate culture, and the economy that will impact your company?

   Fill in: _____

   _____

   _____

Now compress your answers to the preceding six questions into one statement that would represent a realistic strategic direction for your company, business unit, or product.* Also, as you develop your strategic direction, recognize that your corporate culture is the operating system and nerve center of your organization. Therefore, make certain that your company's culture can support your vision. (See Strategy Rule 6: Align Competitive Strategy with Your Corporate Culture.)

Fill in: _____

_____

_____

Section 2: Objectives and Goals

Planning Guidelines

State your objectives and goals both quantitatively and nonquantitatively (the second box in the top row in Figure A.1). For your primary guideline use a timeframe of three to five years. That is, look again at how you define your strategic direction so that you can develop objectives that will have the broadest impact on the growth of your business.

---

* The following statement is an actual example of a well-written strategic direction: "Our strategic direction is to meet the needs of consumers and healthcare providers for drug-delivery devices by offering a full line of hypodermic products and product systems. Our leadership position will be maintained through internal research and development, licensing of technology, and/or acquisition to provide alternative administration and monitoring systems." The company's previous statement indicated a narrower and more restrictive vision: "Our position is to be a leader in the manufacture of hypodermic needles."

Quantitative Objectives

Indicate major performance expectations such as sales growth, market share, return on investment, profit, and other quantitative objectives required by your management. With the longer timeframe, your objectives are generally broad and relate to the total business or to a few major market segments. (In the tactical section these objectives will be more specific for each product and market.)

Fill in: _____

_____

_____

Nonquantitative Objectives

Think of these objectives as a foundation from which to build onto your organization's existing strengths or core competencies or to eliminate any internal weaknesses. Use the following examples to trigger objectives for your business. Above all, keep your objectives specific, actionable, realistic, and focused on achieving a sustainable competitive advantage.

- Improve supply-chain relationships
- Expand secondary distribution or enhance your product's position on the supply chain
- Build specialty products for market penetration
- Establish or improve competitive intelligence procedures
- Focus training actions on improving skills, discipline, and performance of employees
- Launch new and reposition old products
- Upgrade customer and technical services
- Improve marketing mix (product, price, promotion, and distribution) management

Fill in: _____

_____

_____

Section 3: Growth Strategies

Planning Guidelines

This section outlines the process you can use to secure your objectives and goals. Think of strategies as actions to achieve your longer term objectives and tactics as actions to achieve shorter term objectives. Since this timeframe covers three to five years, strategies are indicated here. The one-year portion, illustrated later in the plan, identifies tactics.

In practice, where you have developed broad-based, long-term objectives, you should list multiple strategies for achieving each objective. In instances where you find it difficult to apply specific strategies, it is appropriate to use general strategy statements.

Suggestion: How you write your strategies can vary according to your individual or team's style. For example, you can restate each objective from section 2, followed by a listing of corresponding strategies. Still another option is to write a general strategy statement followed by a detailed listing of specific objectives and strategies. The key point is that each objective (what you want to accomplish) is followed by one or more strategies (actions to reach your objective).

Overall, your thinking about strategies should include actions related to the following categories:

- Growth, mature, and declining markets
- Brand development and product positioning
- Product quality and value-added options
- Market share expansion
- Supply chain options
- Product, price, and promotion mix
- Asset allocations
- Specific marketing, sales, technology, and manufacturing strengths to be exploited
- Employee training and development
- Corporate culture and internal operating systems

Fill in: _____
_____
_____

Section 4: Business Portfolio Plan

Planning Guidelines

The business portfolio includes listings of existing products and markets and new products and markets. Following a logical progression, it is based on the strategic direction, objectives and goals, and growth strategies outlined in the previous sections.

Suggestion: The content of your portfolio should mirror your strategic direction. That is, the broader the scope of your strategic direction is, the more expansive the range of products and markets in the portfolio should be. Conversely, the narrower the dimension of your strategic direction is, the more limited should be the content of products and markets.

Use the following format and guidelines (see Figure A.2) to develop your own business portfolio:

- Existing products in existing markets (market penetration)
  List those existing products you currently offer to existing customers or market segments. In an appendix of the SBP, you can document in numerical or graphic form sales, profits, and market share data. From such information you can determine if your level of penetration is adequate and if possibilities exist for further growth. After identifying new opportunities, it may be necessary for you to revisit section 3 ("Growth Strategies") and list actions you would take to implement the opportunities.
  Fill in: _____
  _____
  _____

- New products in existing markets (product development)
  In this section, extend your thinking and list potential new products you can offer to existing markets. Again, recall the guideline that the broader the dimension of your strategic direction, the broader the possibilities for the content of your portfolio. Also, you should be thinking in a timeframe of three to five years.
  Fill in: _____
  _____
  _____

- Existing products in new markets (market development)
  Now list your existing products into new markets. Explore possibilities for market development by identifying emerging, neglected, or poorly served segments in which existing products can be utilized.
  Fill in: _____
  _____
  _____

- New products in new markets (diversification)
  This portion of the business portfolio is visionary, since it involves developing new products to meet the needs of new and yet untapped markets. Consider new technologies, global markets, and potential strategic alliances to provide input into this section. Once again, interpret your strategic direction in its broadest context. Do not seek diversification for its own sake. Rather, the

	EXISTING PRODUCTS	NEW PRODUCTS
EXISTING MARKETS	Market Penetration	Product Development
NEW MARKETS	Market Development	Diversification

Figure A.2 Business portfolio plan.

whole purpose of the exercise is for you to develop an organized framework for meaningful expansion.

Fill in: _____

_____

_____

The grid in Figure A.2 is a useful format to fill in your business portfolio of products and markets, both existing and new. The business portfolio completes the strategic portion of your SBP. Now you are ready to proceed to the one-year tactical plan.

Overview of the Strategic Business Plan Tactical Section

The tactical plan, the second row of boxes designated as sections 5 through 9 in Figure A.3, is not a stand-alone plan. It is an integral part of the total SBP.

Where commonalties exist between products and markets, one tactical plan can work as long as you make the appropriate changes in such areas as the sales force and the communications mix (advertising, sales promotion, Internet, and publicity). Where you face substantial differences in the character of your product and markets, develop separate tactical plans.

Suggestion: Avoid the temptation to develop a plan for a business, division, or product line by jumping into the middle of the SBP and beginning the process with the tactical 1-year tactical plan. There are no suitable short cuts because input to the tactical plan flows from two directions: (1) from the strategic portion of the SBP (top row) containing the strategic direction, objectives, strategies, and business portfolio, and (2) from the situation analysis (second row), which progresses to opportunities, annual objectives, tactics, and budgets. Also, the thought process that went into the strategic portion of the plan now flows down to feed the shorter term, action-oriented tactical plan.

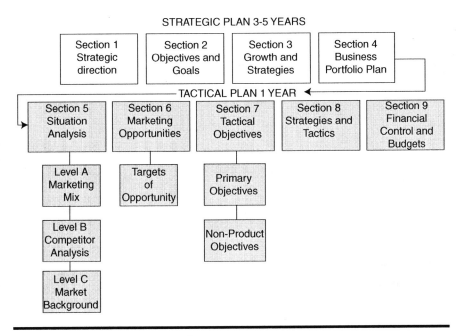

Figure A.3 Strategic business plan.

Section 5: Situation Analysis

The following three-part situation analysis details the past and current situations of your business:

Level A: marketing mix (product, price, supply chain, and promotion)
Level B: competitor analysis
Level C: market background

The purpose of the situation analysis is to define your business in a factual and objective manner. Compile historical data for a period of at least three years. Doing so provides an excellent perspective about where your company has been, where it is now, and where you want it to go as defined in your strategic direction (section 1).

Planning Guidelines

Level A: Marketing Mix—Product

- Objectively describe the performance of your product or service by sales history, profitability, share of market, and other required financial data. Where appropriate, graphically chart sales history with spreadsheets or your company's forms.

Fill in: _____

_____

_____

- Describe your company's current position in the industry related to market share, reputation, product life cycle (introduction, growth, maturity, or decline), and competition.

Fill in: _____

_____

_____

- Describe future trends related to environment, industry, customer, and competitive factors that may affect the position of your product.

Fill in: _____

_____

_____

- Describe the intended purpose of your product in terms of its application or uniqueness.

Fill in: _____

_____

_____

- What are features and benefits of your product as related to quality, performance, safety, convenience, or other factors important to customers?

Fill in: _____

_____

_____

- Describe other pertinent product information, such as expected product improvements and additional product characteristics (size, model, price, packaging); recent features that enhance the position of your product; competitive trends in features, benefits, and technological changes; and changes that would add superior value to the product and provide a competitive advantage.

Fill in: _____

_____

_____

Level A: Marketing Mix—Pricing

- History of pricing
  Examine the history of pricing policies and strategies for each market segment and distribution channel. Consider their impact on the market position of your product.
  Fill in: _____
  _____
  _____

- Future pricing trends
  Describe product pricing trends as they pertain to product specification changes (including formulation and design), financial constraints, and expected market changes (trade and consumer attitudes and competitive responses to price changes).
  Fill in: _____
  _____
  _____

Level A: Marketing Mix—Supply Chain and Methods

- Current channels
  Describe the makeup of your supply chain. Identify the functions performed at each stage within the network (distributor, dealer, direct, e-commerce). Indicate levels of performance (sales volume, profitability, and percentage of business increases). Where appropriate, analyze your physical distribution system, such as warehouse locations, inventory systems, or just-in-time delivery procedures.
  Fill in: _____
  _____
  _____

- Effectiveness of coverage
  Characterize the effectiveness of coverage by the programs and services provided for each channel. Comment on effectiveness of the distribution network (distributors, dealers, direct). Specify the key activities performed at each point and indicate any areas that require corrective action. Also, comment on the impact of future trends in supply chain management, including e-commerce.
  Fill in: _____
  _____
  _____

- **Special functions**
  Indicate special functions performed by your company's sales force for a particular distribution channel and what effect it had on targeted market segments. Also include your distributors' sales forces, if applicable. Comment, too, on such approaches as "push" (through distributors) or "pull" (through end users) strategies.
  Fill in: _____

  _____

  _____

- **Target accounts**
  List target accounts and their level of performance related to sales and quantity. Add comments related to special needs of any customer.
  Fill in: _____

  _____

  _____

- **Future trends**
  Indicate future trends in distribution methods and channels. Project what growth is expected in each major market segment. Also identify how this growth will affect your need for a modified supply chain or method of physical distribution.
  Fill in: _____

  _____

  _____

Level A: Marketing Mix—Advertising, Sales Promotion, Internet, and Publicity

- Analyze your efforts directed at each market segment or distribution channel based on the following elements: expenditures, creative strategy, media, types of promotion, and other forms of communications unique to your industry.
  Fill in: _____

  _____

- **Competitive trends**
  Identify and evaluate competitive trends in the same promotional categories as above. Your advertising agency (or advertising department) and the sales force may prove helpful in compiling this information.

Fill in: _____
_____
_____

- **Strategies**
  Identify your company's past and current communications strategies by product and market segment and describe trends in these areas.
  Fill in: _____
  _____
  _____

- **Other support strategies**
  Identify other support programs (publicity, educational, professional, trade shows, literature, films and videos, the Internet) that you have used and evaluate their effectiveness.
  Fill in: _____
  _____
  _____

Level B: Competitive Analysis

- **Market share**
  List all your competitors in descending-size order along with their sales and market shares. Include your company's ranking within the listing. Show at least three competitors—more if the information is meaningful. (If unable to provide useable information for this portion and the others that follow, you should give very high priority to developing a business intelligence initiative, of which competitor intelligence is a key component.)
  Fill in: _____
  _____
  _____

- **Competitors' strengths and weaknesses**
  Identify each competitor's strengths and weaknesses related to such factors as product quality, distribution, pricing, promotion, management leadership, and financial condition. Also, indicate any significant trends that would signal unsettling market situations, such as aggressiveness in growing market share or excessive discounting to maintain market position. Attempt to make your competitive analysis as comprehensive as possible. The more competitive intelligence you gather, the more strategy options you have open to you.

(To assist you in developing a quality analysis, review the strategy rules in this book, with particular attention to the strategy diagnostic tools, to determine how closely your competitors adhere to them. You will thereby secure a more comprehensive picture of your competitors.)

Fill in: _____

_____

- Product competitiveness

    Identify competitive pricing strategies, price lines, and price discounts, if any. Identify those competitors firmly entrenched in low-price segments of the market, those at the high end of the market, or competitors that are low-cost producers.

    Fill in: _____

    _____

- Product features and benefits

    Compare the specific product features and benefits with those of competitive products. In particular, focus on product quality, design factors, and performance. Evaluate price and value relationships for each, discuss customer preferences (if available), and identify unique product innovations.

    Fill in: _____

    _____

- Advertising effectiveness

    Identify competitive spending levels and their effectiveness, as measured by awareness levels, competitive copy test scores, and reach and frequency levels (if available). Such measurements are conducted through formal advertising research conducted by your advertising agency, independent marketing research firms, or some publications. Where no reliable quantitative research exists, use informal observation or rough measurements of advertising frequency and type.

    Fill in: _____

    _____

- Effectiveness of the supply chain

    Compare competitive distribution strengths and weaknesses. Address differences in market penetration, market coverage, delivery time, and physical movement of the product by regions or territories. Also identify major accounts where competitors' sales are weak or strong.

Fill in: _____

_____

- Packaging
  Compare the package performance, innovation, and preference of competitive products. Also review size, shape, function, convenience of handling, ease of storage, and shipping.

  Fill in: _____

  _____

- Trade and consumer attitudes
  Review both trade (distributor or dealer) and consumer attitudes toward product quality, customer and technical service, company image, and company performance.

  Fill in: _____

  _____

- Competitive share of market trends
  While share of market was previously included as a way of determining overall performance, the intent here is to specify trends in market share gains by individual products, as well as by market segments. Further, identify where each competitor is making a major commitment and where it may be relinquishing control by product and segment.

  Fill in: _____

  _____

- Sales force effectiveness and market coverage
  Review effectiveness as it relates to sales, service, frequency of contact, and problem-solving capabilities by competitor and by market segment. Look to all sales force performance within the supply chain. For example, if you are a manufacturer, look at your distributors' market coverage. Then examine distributors' coverage of their customers, which could be dealers or end users.

  Fill in: _____

  _____

Level C: Market Background

This last part of the situation analysis focuses on the demographic and behavioral factors of your market. Doing so helps you determine market size and customer preferences (both business-to-business and business-to-consumer) in a changing competitive environment.

You can derive data from primary market research (market segmentation studies, awareness levels, and product usage studies) or from secondary sources (trade and government reports). If you give careful attention to compiling accurate information, you will benefit from reliable input for developing sections 6, 7, and 8 of the SBP. This information also highlights any gaps in knowledge about markets and customers and thereby helps you determine what additional market intelligence is needed to make more effective decisions. The following categories are considered part of the market background:

- Customer profile
  Define the profile of present and potential end-use customers that you or your distributors serve. Your intent is to look further down the distribution chain and view the end-use consumer. Examine market segments, distributors, and dealers served. Address this factor from your distributors' point of view.
  Fill in: _____

- Distributors' overall sales
  Concentrate on classifying the key customers that represent the majority of sales.
  Fill in: _____

- Other classifications
  Profile your customers by such additional factors as types of products used, level of sophistication, price sensitivity, and service. Also indicate any target accounts that you can reach directly, thereby bypassing the distributor.
  Fill in: _____

- Frequency and magnitude of products used
  Define customer purchases by frequency, volume, and seasonality of purchase. Additional information might include customer inventory levels,

retail-stocking policies, and volume discounts. Also look at consumer buying behavior related to price, point-of-purchase influences, or coupons.

Fill in: _____

_____

- Geographic aspects of products used
  Define customer purchases regionally or territorially (both trade and consumer). Segment buyers by specific geographic area (e.g., rural, urban) or by other factors relevant to your business.

  Fill in: _____

  _____

- Market characteristics
  Assess the demographic, psychographic (lifestyle), and other relevant characteristics of your customers. Also examine levels of product technology in use; purchase patterns and any distinctive individual or group behavioral styles; and attitudes toward the company's products, services, quality, and image.

  Fill in: _____

  _____

- Decision maker
  Define who makes the buying decisions and when and where they are made. Note the various individuals or departments that many influence the decision.

  Fill in: _____

  _____

- Customer motivations
  Identify the key motivations that drive your customers to buy the product. Why do they select one manufacturer (or service provider) over another? Customers may buy your product because of quality, performance, image, technical or customer service, convenience, location, delivery, access to upper-level management, friendship, or peer pressure.

  Fill in: _____

  _____

- Customer awareness
  Define the level of consumer awareness of your products. To what extent do consumers:
  - Recognize a need for your type of product?
  - Identify your product, brand, or company as a possible supplier?
  - Associate your product, brand, or company with desirable features?

  Fill in: _____

  _____

  _____

- Segment trends
  Define the trends in the size and character of the various segments or niches. (A segment is a portion of an entire market; a niche is part of a segment.) A segment should be considered if it is accessible, measurable, potentially profitable, and has long-term growth potential. Segmenting a market also serves as an offensive strategy to identify emerging, neglected, or poorly served markets that can catapult you to further sales growth. You can also consider segments as a defensive strategy to prevent inroads of a potential competitor through an unattended market segment.

  Fill in: _____

  _____

  _____

- Other comments and critical issues
  Add general comments that expand your knowledge of the market and customer base. Also identify any critical issues that have surfaced as a result of conducting the situation analysis that should be singled out for special attention.

  Fill in: _____

  _____

  _____

Section 6: Marketing Opportunities

Planning Guidelines

In this section, you examine marketing strengths, weaknesses, and options. Opportunities will begin to emerge as you consider the variety of alternatives. Try to avoid restricted thinking. Take your time and brainstorm. Dig for opportunities with other members of your planning team. If one does not exist, then put together a team representing different functional areas of the business (or persuade senior management to approve its formation).

Consider all possibilities for expanding existing market coverage and laying the groundwork for entering new markets. Also consider opportunities related to your competition. For instance, offensively, which competitors can you displace from which market segments? Defensively, which competitors can you deny entry into your market?

As you go through this section, revisit your strategic portion of the SBP (top row of boxes in Figure A.1). While that portion represents a three- to five-year period, work must begin at some point to activate the strategic direction, objectives, growth strategies, and business portfolio sections. Further, refer to the situation analysis in the last section—specifically, the competitive analysis—for voids or weaknesses that could represent opportunities. Note the two-directional flow used to create opportunities: (1) The visionary thinking you used to shape the strategic portion of the SBP now flows down to focus on one-year opportunities, and (2) the situation analysis that exposes voids and weaknesses also represents opportunities.

Now review the following screening process to identify your major opportunities and challenges. Once you identify and prioritize the opportunities, convert them into objectives and tactics, which are the topics of the next two sections of the SBP.

- Present markets
  Identify the best opportunities for expanding present markets through:
  - Cultivating a new revenue stream through new users
  - Displacing competition
  - Increasing product usage or services by present customers
  - Redefining market segments
  - Reformulating or repackaging the product
  - Identifying new uses (applications) for the product
  - Repositioning the product to create a more favorable perception by consumers and by developing a competitive advantage over rival products
  - Expanding into new or unserved market niches

  Fill in: _____

  _____

  _____

- Targets of opportunity
  List any areas outside your current market segment or product line not included in the above categories that you would like to explore. Be innovative and entrepreneurial in your thinking. These areas are opportunistic. Therefore, due to their innovative and risky characteristics, they are isolated from the other opportunities. Those you select for special attention are placed in a separate part of the objectives section of the SBP.

  Fill in: _____

  _____

  _____

Section 7: Tactical Objectives

At this point, you have reported relevant factual data in section 5; you have interpreted their meaning and business-building potential of your product line in section 6. You must now set the objectives you want to achieve during the current planning cycle—generally defined as a 12-month period to correspond with annual budgeting procedures.

Once again, you will find it useful to review sections 5 and 6. Also, it will be helpful to review the strategic portion of the plan. You want to be certain that actions related to your long-range strategic direction, objectives, and strategies are incorporated into your tactical one-year objectives.

This section consists of three parts:

- Assumptions: projections about future conditions and trends
- Primary objectives: quantitative measurements related to your responsibility, including targets of opportunity
- Functional objectives: operational goals for various parts of the business

Planning Guidelines

Assumptions

For objectives to be realistic and achievable, you must first generate assumptions and projections about future conditions and trends. List only those major assumptions that will affect your business for the planning year as it relates to the following:

- Economic assumptions: comment on the overall economy, local market economies, industrial production, plant and equipment expenditures, consumer expenditures, and changes in customer needs. Also document market size, growth rate, costs, and trends in major market segments.

  Fill in:_____

  _____

- Technological assumptions: include depth of research and development efforts, likelihood of technological breakthroughs, availability of raw materials, and plant capacity.

  Fill in:_____

  _____

- Sociopolitical assumptions: indicate prospective legislation, political tensions, tax outlook, population patterns, educational factors, and changes in customer habits.
  Fill in: _____

- Competitive assumptions: identify activities of existing competitors, inroads of new competitors, with particular attention to surging global competitors from Asia and Eastern Europe, and changes in trade practices.
  Fill in: _____

Primary Objectives

Focus on the primary financial objectives that your organization requires. Also include targets of opportunity that you initially identified as innovative and entrepreneurial in section 6. When there are multiple objectives, you may find it helpful to rank them in priority order. Be sure to quantify expected results where possible. You can separate your objectives into (1) primary objectives, which are current and projected sales, profits, market share, return on investment, and other quantitative measures (use Exhibit A.1, a form provided by your organization, or any spreadsheet software), and (2) targets of opportunity objectives, which include functional objectives in which you state the objectives relating to both product and nonproduct issues in each of the following categories. (You can alter the list of objectives to fit your business and industry.)

Exhibit A.1   Primary Objectives

Product Group Breakdown	Current				Projected			
	Sales ($)	Units	Margins	Share of Market	Sales ($)	Units	Margins	Share of Market
Product A								
Product B								
Product C								
Product D								

Targets of Opportunity Objectives

Product Objectives

- Quality

  Identify quality objectives that would achieve a competitive advantage by exceeding industry standards in some or all segments of your market.

  Fill in: _____

  _____

  _____

- Development

  Deal with new technology through internal R&D, licensing, or joint ventures.

  Fill in: _____

  _____

  _____

- Modification

  Deliver major or minor product changes through reformulation or engineering.

  Fill in: _____

  _____

  _____

- Differentiation

  Enhance competitive position through function, design, or any other changes that can differentiate a product or service.

  Fill in: _____

  _____

  _____

- Diversification

  Transfer technology, use the actual product in new applications, or diversify into new geographic areas, such as developing countries.

  Fill in: _____

  _____

  _____

- Deletion

  Remove a product from the line due to unsatisfactory performance, or keep it in the line if the product serves some strategic purpose, such as presenting your company to the market as a full-line supplier.

Fill in: _____
_____
_____

- Segmentation
  Create line extensions (adding product varieties) to reach new market niches or defend against an incoming competitor in an existing market segment.
  Fill in: _____
  _____
  _____

- Pricing
  Include list prices, volume discounts, and promotional rebates.
  Fill in: _____
  _____
  _____

- Promotion
  Develop sales force support, sales promotion, advertising, Internet, and publicity to the trade and consumers.
  Fill in: _____
  _____
  _____

- Supply chain
  Add new distributors to increase geographic coverage, develop programs or services to solidify relationships with the trade, remove distributors or dealers from the chain, or maintain direct contact with the end user.
  Fill in: _____
  _____
  _____

- Physical distribution
  Identify logistical factors that would include order entry to the physical movement of a product through the supply chain and eventual delivery to the end user.
  Fill in: _____
  _____
  _____

- Packaging
  Use functional design and decorative considerations for brand identification.
  Fill in:_____
  _____
  _____

- Service
  Broaden the range of services, from providing customers access to key executives in your firm to providing on-site technical assistance.
  Fill in:_____
  _____
  _____

- Other
  Indicate other objectives, as suggested in "Targets of opportunities."

Nonproduct Objectives

Although most activities eventually relate to the product or service, some are support functions that you may or may not influence. How much influence you can exert depends on the functions represented on your planning team.

- Target accounts
  Indicate those customers with whom you can develop special relationships through customized products, distribution or warehousing, value-added services, or participation in quality improvement programs.
  Fill in:_____
  _____
  _____

- Manufacturing
  Identify special activities that would provide a competitive advantage, such as offering small production runs to accommodate the changing needs of customers and reduce inventory levels.
  Fill in:_____
  _____
  _____

- Marketing research
  Cite any customer studies that identify key buying factors and include competitive intelligence.

Fill in: _____
_____
_____

- Credit
  Include any programs that use credit and finance as a value-added component for a product offering, such as rendering financial advice or providing financial assistance to customers in certain situations.
  Fill in: _____
  _____
  _____

- Technical sales activities
  Include any support activities, such as 24/7 hot-line telephone assistance that offers on-site consultation to solve customers' problems.
  Fill in: _____
  _____
  _____

- R&D
  Indicate internal research and development projects as well as joint ventures that would complement the strategic direction identified in section 1 of the SBP.
  Fill in: _____
  _____
  _____

- Training
  List internal training programs, as well as external end-user programs.
  Fill in: _____
  _____
  _____

- Human resource development
  Identify specialized skills required by those individuals who would make the SBP operational.
  Fill in: _____
  _____
  _____

- **Other**
  Include specialized activities that may be unique to your organization.
  Fill in: _____
  _____
  _____

Section 8: Strategies and Tactics

Strategy is the art of coordinating the means (money, human resources, materials) to achieve the ends (profits, customer satisfaction, growth) as defined by company policy, strategic direction, and objectives. From another perspective, strategies are actions to achieve long-term objectives; tactics are actions to achieve short-term objectives.

Therefore, in this section, strategies and tactics have to be identified and put into action. Responsibilities are assigned, schedules set, budgets established, and checkpoints determined. Make sure that the members of the planning team actively participate in this section. They are the ones who have to implement the strategies and tactics.

Planning Guidelines

- Restate the functional product and nonproduct objectives from section 7 and link them to the strategies and tactics you will use to reach each objective. One of the reasons for doing this is to clarify the frequent misunderstanding between objectives and strategies. Objectives are what you want to accomplish; strategies are actions that indicate how you intend to achieve your objectives. (Note: If you state an objective and do not have a related strategy, you may not have an objective. Instead, the statement may be an action for some other objective.)
  Fill in: _____
  _____
  _____

  Objective 1: _____
      Strategy/tactic _____
  Objective 2: _____
      Strategy/tactic _____
  Objective 3: _____
      Strategy/tactic _____

Summary Strategy

Summarize the basic strategies for achieving your primary objectives. Also, include alternative and contingency plans if situations should arise to prevent you from reaching your objectives. Be certain, however, that such alternatives relate to the overall SBP. As you develop your final strategy statement, use the following checklist to determine its completeness:

- Changes needed to the product or package, including differentiation and value-added features
- Strategies to create a competitive advantage, along with contingency plans to counter competitors' aggressive moves
- Changes to price, discounts, or long-term contracts to address market share issues
- Changes to advertising strategy, such as the selection of features and benefits or copy themes to special groups
- Strategies to reach new, poorly served, or unserved market segments
- Promotion strategies related to private-label products; dealer and distributor, consumer, and sales force incentives
- Internal changes in the operating systems, as well as initiatives that would better align the corporate culture with the strategies

Fill in: _____

_____

_____

Section 9: Financial Controls and Budgets

Planning Guidelines

Having completed the strategy phase of your SBP, you must decide how you will monitor its execution. Therefore, before implementing it, you have to develop procedures for both control (comparing actual and planned figures) and review (deciding whether planned figures should be adjusted or other corrective measures taken).

This final section incorporates your operating budget. If your organization has reporting procedures, you should incorporate them within this section. Included here are examples of additional reports or data sheets designed to monitor progress at key checkpoints of the plan and to permit either major shifts in strategies or simple midcourse corrections:

- Forecast models
- Sales by channel of distribution
    - Inventory or out-of-stock reports
    - Average selling price (including discounts, rebates, or allowances) along the supply chain and by customer outlet
- Profit and loss statements by product
- Direct product budgets
- R&D expenses
- Administrative budget
- Spending by quarter

Regardless of the forms you use, as an overall guideline make certain that the system serves as a reliable feedback mechanism. Your interest is in maintaining explicit and timely control so that you can react swiftly to impending problems. Further, it should serve as a procedure for reviewing schedules and strategies.

Finally, the system should provide an upward flow of fresh market intelligence that, in turn, could impact broad policy revisions at the highest levels of the organization.

The only other part left in your SBP is an appendix. It should include pertinent industry data and market research; additional data on competitors' strategies, including information on their products, pricing, promotion, distribution, and profiles of management leadership (if available); and details about new product features and benefits. (Various computer-based databases, CRM, and other software programs can assist to strengthen your plan.)

Index

A

Abercombie & Fitch Co. (A&F), 118
Accounting personnel, role in establishing security, 31
Action(s)
 estimates and calculations in, 4–6
 prioritizing, 8–10
 successful leaders and, 135
Advanced Micro Devices (AMD), 103–104
Advertising personnel, role in establishing security, 31
Agents
 double, 100
 employing, 98
 expendable, 100
 inside, 99
 living, 100–101
 native, 98–99
Alliances, company operations and, 42
Amazon.com, 175
Ambition, leadership and, 137, 138–139
American Express, 29
Anticipation, morale and, 163
Apple Computer, Inc.
 indirect strategy applications at, 21
 iPod case, 79
 leadership at, 140–142
 market position of, 63–64
 shift to the offensive strategy at, 3
 Sony and, 115–116
 strategy of, 74
 use of ethnographic studies, 67
Arrogance, speed and, 52
Arrow Electronics, 11
Audacity, competition and, 3

B

Baby boomers, marketers and, 60
Bank of America, 75
Beliefs, in corporate culture, 109, 119–122
Belonging needs, in Maslow's hierarchy of needs, 153
Benchmarking, 94
Best Buy, 24–25
BMW, 64
Boeing Co., 118–119, 155–157
Bold manager, 6
Boldness
 energizing company culture with, 124
 shifting to the offensive with, 3–4
BorgWarner, 132
BPM system, *see* Business performance management (BPM) system
Business culture, *see* Corporate culture
Business history, valuing, 176–178
Business performance management (BPM) system, 96
Business plans; *see also* Strategic business plan (SBP)
 interface of SWOT and, 9
 policy and, 7–8
 poststrategy and, 33–36
Business portfolio plan, 194–196
Business strategy; *see also* Strategy(ies)
 aligning with corporate culture, 116–117
 of Apple Computer, 141
 competitive intelligence and, 89–92
 data mining and, 29
 impact of morale on, 158–159
 leadership and, 109
 powering up, 143–144

C

Candid attitude, in building morale, 164
Canon, 30, 72
Capital Financial Corp., 161
Cargill, 120
Caterpillar Inc., 32
Caution, impact on competition, 3
Central markets, 75–76, 80
Chains of command, as barrier to speed, 49–50
Challenging markets, 76, 80
Chambers, John T., 121
Character, leadership and, 137, 138
Charles Schwab & Co., 29–30, 121–123
China, concentration strategy and, 58
Circuit City, 24–25
Cisco Systems, 42, 121
Citigroup, 47, 75
Coca-Cola, 82
Communication(s)
　building morale and, 163
　speed and, 51
Communications skills, morale and, 162–163
Company(ies)
　sampling of actual problems, 173
　speed and, 39
Compassionate manager, 5
Competition
　moving to the offense with, 3
　selecting market segment and shaping, 64–67
Competitive benchmarking, competitive intelligence and, 94–95
Competitive intelligence (CI); *see also* Internal competitive intelligence
　applications of, 101–104
　business strategy and, 89–92
　gathering information for, 28–30
　introduction to, 87–89
　questions on, 91–92
　six-step process, 101
　strategy diagnostic tool, 105–106
　taking action and, 4
　tools and techniques for, 92–95
Competitive Strategy, Techniques for Analyzing Industries and Competitors (Porter), 14
Competitors; *see also* Competition
　aggressive, 9, 50–52, 125
　decoding, 117–123
　direct strategy and, 21–22
　strengthening security and, 31
　tips for assessing, 11–12
　using direct strategy against, 21–22
　using indirect strategy against, 20
Complacency, speed and, 52
Computer Associates (CA), 133–134
Concentration strategy
　applications of, 61–63
　introduction to, 57–64
　market categories as guide to, 69–82
　for market segments, 64–67
　rules of, 65
　strategy diagnostic tool, 83–85
　utilizing grassroots ethnography and, 67–69
Conceptual skills, leaders and, 139
Consumer behavior, 12
Corporate culture
　aligning business strategies with, 116–117
　business practice and, 113
　decoding competitors and, 117–123
　energizing, 124–126
　highly conservative, speed and, 49
　high-performing, characteristics of, 108–116
　introduction to, 107–110
　of Nucor, 175
　strategic diagnostic tool, 127–128
　strategy applications in, 61
Courage, 46, 142
Creativity
　assessing, 12–14
　energizing corporate culture with, 124
Creativity economy, 12
Culture, speed and, 51–52
Customer relationships, timing and, 40
Customers, enhancing relationships with, 33, 35–36
Customer surveys, competitive intelligence and, 92

D

Dashboards, 96
Dassault Systems, 49
Databases, searching competitors with, 93
Data mining, 29
Decision-making
　activating intuition and, 179–183
　human element and, 184–185
　introduction to, 171–176
　tips for, 183–184

managing knowledge and, 178–179
strategy diagnostic tool, 186–187
valuing business history and, 176–178
Defensive action, using, 1–2
Delegating, in building morale, 163
Dell, Inc.
    customer support competition and, 63–64
    EMC and, 42
    indirect strategy of, 21
    leadership at, 138
    memory chips of, 75
    supply chain management, 76
DHL, 36
Difficult markets, 76–77, 80
Direction, successful leaders and, 135
Direct leadership, 135–136
Direct strategy, competitors and, 21–22
Discipline
    of employees, speed and, 51
    leadership and, 137
Diversification of products, 35
Diversity, corporate culture and, 126
Double agents, 100
Drive, morale and, 162

E

eBay, 46, 175
Efficiency, integrating speed with, 42
Electrolux, 137–138
EMC Corp., 42, 52
Emerging markets, 42
Employee(s); *see also* Personnel
    as barrier to speed, 47, 51
    building morale of, 164
    corporate culture and, 109, 126
    elevating morale of, 42
Encircled markets, 77–79, 80, 81
Enron Corp., 112, 113, 139
Enthusiasm, corporate culture and, 126
Entrepreneurship
    corporate culture and, 108
    of Nucor, 175–176
Environment
    competitive intelligence techniques and, 92–101
    energizing company culture and, 124–125
    flexible, living in, 124–125

Erratic behavior, morale and, 163
Esteem needs, in Maslow's hierarchy of needs, 154
Estimates and calculations, in decisive action, 4–8
Ethnography tool, 67
Executives, boldness and courage of, 2
Expendable agents, 100
Expertise, within organization, 22–23
Explicit knowledge, 178
Exploit process, in maneuvering, 103
External relationships, estimating, 4–6
Extraordinary activities, 26–27

F

Fairness, corporate culture and, 108
FedEx, 36
Financial controls and budgets, in strategic business plan, 215–216
Financial resources, strategic goals and, 26–27
Fiorina, Carleton, 48
Fisher-Price, 78
Flextronics, 11
Ford Motor Co., 32
Friction
    evaluating ability to deal with, 14–15
    limiting negative effects of, 15–16

G

Gap, Inc., 13
Gateway, 63–64
GE Aircraft Engineering Division, 123
General Electric Co., 12, 64, 111–112, 162
General Motors (GM), 5, 52, 58, 64
Google Inc., 46, 65, 131–132, 162
Government agencies, competitive intelligence and, 93
Grassroots ethnography guidelines
    creating special language in, 68
    describing rituals in, 68–69
    map a segment in, 67
    observing body language, 68
Growth strategies, in strategic business plan, 193–194

H

Haier Group, 59
Heineken Brewery, 123
Herzberg's motivation-hygiene theory, 152
Heublein, 28–29
Hewlett-Packard (H-P), 47–48, 52, 63–64, 72, 182
Hilton Hotels, 30
Hoechst, 136–137
Hold reserves, exploiting opportunities to, 10–12
Home Depot Inc., 150, 162
Human element, in decision making, 184–185
Hurd, Mark V, 47
Hyundai Motor Inc., 7, 42

I

IBM, 21, 64, 161
Image, customers and, 35
Immelt, Jeffrey, 111
Indecisiveness, managers and, 46
Independence, corporate culture and, 108
Indirect strategy
    in action, 20–21
    advantages and applications of, 21
    developing, 22–26
    developing poststrategy and, 33–36
    establishing security and, 30–31
    gathering competitive intelligence and, 28–30
    implementing, 32–33
    introduction to, 19–20
    power of, 20–22
    sources for creating, 27
    strategy diagnostic tool, 36–38
Individuals, speed and, 39
Industry studies, competitive intelligence and, 94
Innovation
    assessing, 12–14
    energizing corporate culture with, 124
    interface with morale, 164–165
    technology, for supply chain, 60, 63
In-person interviews, competitive intelligence and, 92–93
Inside agents, 99

Insightful manager, 5
Intel Corp., 41, 67, 90–92
Internal competitive intelligence
    market signals and, 97–98
    in organizations, evaluating, 95–96
    reverse engineering and, 96–97
Internal relationships, estimating, 4–6
Internet, 29
Interpersonal skills, leaders and, 139
Intuit, 123
Intuition
    activating, 179–183
    decision-making and, 171–172
    friction and, 15–16

J

J. P. Morgan Chase, 75
Jobs, Stephen, 3, 140
Johnson & Johnson (J&J), 160–161, 163

K

Kellogg, 64
Key markets, 73, 80
Kikukawa, Tsuyoshi, 2
Knowledge economy, connecting to, 12
Knowledge management, decision-making and, 178–179
Knowledge skill, successful leaders and, 135
Kodak, 30
Kotler, Philip, 117

L

Laid-back competitor, 117
Leadership
    in competitive arena, 140–144
    in corporate culture, 109
    in estimating and calculating action, 5
    indirect strategy and, 26–27
    introduction to, 129–135
    levels of, 135–139
    mediocre, speed and, 45

role of, 129–135
skills of, 139–140
strategy diagnostic tool, 145–147
Leading-edge markets, 72–73, 80
LEGO Group, 120
LG. Philips, 52
Life cycle of products, extending, 34–35
Line extension of products, 34
Linked markets, 73–75, 80
Linux, 65–66
Living agents, 100–101
Liz Claiborne Inc., 77
Lutz, Robert, 5

M

Mail interview, competitive intelligence and, 93
Management control, at Apple Computer, 141
Managerial competence, in corporate culture, 109
Managers
  as barrier to speed, 46, 47
  line, speed and, 48–49
  quality of, 5–6
Manganello, Timothy, 132
Mapping market segment, 67
Market allegiance, 70
Market categories
  central markets, 75–76, 80
  challenging markets, 76, 80
  difficult markets, 76–77, 80
  encircled markets, 77–79, 80, 81
  key markets, 73, 80
  leading edge markets, 72–73, 80
  linked markets, 73–75, 80
  natural markets, 69
Marketing, 71
Marketing Management (Kotler), 117
Marketing opportunities, in strategic business plan, 206–207
Marketing/promotion, concentration strategy applications in, 62
Market intelligence, speed and, 44–45
Market position
  building solid, 125
  customers and, 35
Market research personnel, role in establishing security, 31

Market segment
  applying body language for defining, 68
  concentrating on, 58–60
  mapping, 67
  market characteristics checklist, 70–71
  rituals and, 68–69
  shaping competition and selecting, 64–67
  using language for describing, 68
Market selection
  competitive intelligence and, 101–102
  in estimating and calculating action, 6–7
Market share, timing and, 40
Market signals, 97–98
Market size, 70
Market uncertainty, decision-making and, 171–176
Marriott International Inc., 82
Maslow's hierarchy of needs, 152–154
Material resources, strategic goals and, 26–27
Mattel, 78
McGregor's X and Y theories, 152, 153
McNerney, W. James Jr., 156
Merck, 151
Microsoft Corporation, 46, 75
  concentration strategy and, 65–66
  leadership at, 138
  managers at, 48–49
  Symantec Corp. and, 119
Miller Brewing, 21, 64
Models creation, in fostering creativity, 13
Modification of products, 34
Momentum, loosing, 40–41
Morale
  action steps to build, 163–164
  aggressive competitors and, 50–52
  barriers to building, 161–163
  human heart and, 157–158
  impact on business strategy, 158–159
  integrating into leadership style, 149–152
  interface with innovation, 164–165
  maintaining momentum and, 159–161
  motivational behavior and, 152–155
  power of unity and, 158
  reasons for drop in, 165
  strategy diagnostic tool, 166–169
  technology and, 155–157
  trust and, 166
Motivation
  Herzberg's factors affecting, 152
  at Nucor, 176
  successful leaders and, 135

Motivational behavior, 152–158
Motorola Inc., 46
    competitors *vs.*, 40–41
    concentration strategy of, 59
    creativity and innovation assessment at, 13–14
    cross-functional teams at, 78
    Nokia and, 13–14, 22, 41

N

Name brand, customers and, 35
Nardelli, Robert, 150–151
Narrative development, in fostering creativity, 13
Native agents, 98–99
Natural markets, 69, 71–72, 80
Nike, 67, 82
Nissan, 184–185
Nokia, 46, 75
    concentration strategy of, 59
    leadership at, 138
    Motorola and, 13–14, 22, 41
    Whirlpool Corp. and, 182
Normal activities, 26
Nucor Corp., 174–176

O

Objectives and goals, in strategic business plan, 192–194
Observation, in fostering creativity, 13
Olympus Corp., 2
On-site observations, competitive intelligence and, 94
Openness, corporate culture and, 108
Operations of customers, enhancing, 35
Organizational layers, speed and, 49–50
Organizational leadership, 136
Organizational structure, in corporate culture and, 109
Organizations
    evaluation of, 96
    flexibility of, market conditions and, 70
Ouchi's theory Z, 154–155

P

Penetrate process, in maneuvering, 102
Performance
    monitoring, morale and, 164
    of Nucor, 175
Personnel, 71; *see also* Employee(s)
    development of, strategy and, 61
    role in establishing security, 31
Physical distribution, concentration strategy applications in, 63
Planning skills, building morale and, 161–162
Policy, in estimating and calculating action, 7–8
Porter, M., 14
Positioning, 70, 103
Positive attitude, building morale and, 164
Poststrategy development, 33–36
Price, concentration strategy applications in, 62
Pricing, 70
Pride, 126, 137
Priorities, 8–10, 163–164
Process installation, in fostering creativity, 13
Procrastination, morale and, 163
Procter & Gamble (P&G)
    alliances of, 42
    concentration strategy of, 59
    employment of Web technologies, 64
    Marriott and, 82
    personnel development at, 64
    shift to creativity economy, 12
    tapping beliefs, 120–121
    use of ethnographic studies, 67
    use of social network analysis, 161
    vision of, 162
    Whirlpool Corp. and, 182
Production
    personnel, role in establishing security, 31
    speeding up, 35
Product life cycle
    prolonging, 33
    short, as barrier to speed, 49
Product(s)
    as commodity, speed and, 41
    concentration strategy applications in, 61
    creating and prolonging life of, 34–35
    ethnographic studies and design of, 67
    extending life of, 36
    life cycle of, 41–42, 49
    phasing out strategy of, 34

position, timing and, 40
  use of, 70
Progress, mediocre leadership and, 46
Psychological effect, indirect strategy and, 20, 29
Psychological needs, in Maslow's hierarchy of needs, 153
Published data, competitive intelligence and, 93
Purpose, successful leaders and, 135

Q

Questions, building morale and, 163

R

Reis, Al, 103
Re-merchandising products, 35
Renault, 184–185
Reordering procedures, examining, 36
Resilience, corporate culture and, 108
Resourcefulness, morale and, 162
Resources, strategic goals and determining, 26–27
Respecting differences, in building morale, 164
Returns and complaints, customers and, 35
Reverse engineering, 96–97
Ricoh Co., 4, 30, 72
Risk, new idea and, 2
Rituals
  impact on corporate culture, 122–123
  market segment description of, 68–69
Royal Dutch Shell, 49

S

Safety needs, in Maslow's hierarchy of needs, 153
Sakurai, Masamitsu, 4
Sales force, competitive intelligence and, 92
Salesforce.com, concentration strategy of, 65–66

Sales manager, role in establishing security, 31
Sales reps, role in establishing security, 31
Salk, Jonas, 179
Samsung Electronics
  concentration strategy of, 58–59
  customer loyalty at, 74–75
  management of, 138
  new strategy of, 14
  speed and, 41, 52
SanDisk Corp., 41
SAP, 49
SBC Communications, 46
Seasonal forces, in estimating and calculating action, 6
Security, establishing, 30–31
Selective competitor, 117
Self-actualization needs, in Maslow's hierarchy of needs, 154
Self-confidence
  leadership and, 134–135
  morale and, 162
Self-development, morale and, 162
Self-esteem, 46
Senior management, as barrier to speed, 48
Services, concentration strategy applications in, 61
Shaheen, George, 6
Sharp Electronics, 52
Sherman, William T., 4
Shift to the offense
  assessing creativity and innovation, 12–14
  assessing friction, 14–16
  with boldness, 3–10
  holding reserves and, 10–12
  introduction to, 1–2
  rule of speed vs., 40
  strategy diagnostic tool, 16–18
Siebel Systems Inc., 6, 40
Simulations, in fostering creativity, 13
Situational analysis, in strategic business plan, 197–206
Social network analysis, 142, 161
Solectron, 11
Sony Corporation, 12, 74, 115–116, 138
Speed strategy
  barriers to, 43–52
  as core rule of strategy, 52
  guiding principles of, 43
  introduction to, 39–40
  strategic value of, 40–43
  strategy diagnostic tool, 53–55

Sprint Corp., 41
Standards, setting high, building morale and, 164
Starbucks, 138, 175
Stochastic competitor, 117
Straberg, Hans, 137
Straightforward manager, 5
Strategic business plan (SBP)
  business portfolio plan in, 194–196
  financial controls and budgets in, 215–216
  growth strategies in, 193–194
  marketing opportunities in, 206–207
  objectives and goals in, 192–194
  overview of, 189
  situational analysis in, 197–206
  strategic direction in, 190–191
  strategies and tactics in, 214–215
  tactical objectives in, 208–214
  tactical section in, 196–197
Strategic direction, in strategic business plan, 190–191
Strategic goals
  areas of expertise in, 22–23
  categories of customers in, 23
  customers' problems resolution in, 24
  determining resources in, 26–27
  external factors in, 24–25
  pinpointing market segments in, 23
  technologies to satisfy customers in, 24
Strategic leadership, 136
Strategy diagnostic tool
  act with speed, 53–55
  aligning competitive strategy with corporate culture, 127–128
  create a morale advantage, 166–169
  develop leadership skills, 145–147
  grow by concentration, 83–85
  maneuver by indirect strategy, 36–38
  prioritize competitive intelligence, 105–106
  shift to the offensive, 16–18
  strengthen your decision-making capabilities, 186–187
Strategy(ies); *see also* Business strategy
  implementing, 32–33
  intuition and, 15–16
  leadership and implementing, 142–143
  for phasing out product, 34
  in strategic business plan, 214–215

Strategy rules
  act with speed, 39–55
  align competitive strategy with your corporate culture, 107–128
  create a morale advantage, 149–169
  develop leadership skills, 129–147
  grow by concentration, 57–80
  maneuver by indirect strategy, 19–38
  prioritize competitive intelligence, 87–106
  shift to the offense, 1–18
  strengthen your decision-making capabilities, 171–187
Strengths, assessing, 65–67
Strengths, weaknesses, opportunities, threats, *see* SWOT analysis
Strict manager, 5
Success, strategies and, 26–27
Supply chain
  concentration strategy applications in, 63
  effectiveness of, 71
  loosing position in, 42
  for technology innovation, 60, 63
Support from senior management, speed and, 48
SWOT analysis
  benefits of, 8–9
  interface of business planning with, 9
Symantec Corp., 119
Symbols, impact on corporate culture, 122–123

T

Tacit knowledge, 178
Tactical objectives, in strategic business plan, 196–197, 208–214
Tactical skills, leaders and, 140
Tactics, 61, 214–215
Teams, functions and responsibilities of, 143
Technical skills, leaders and, 140
Technology
  evolving, staying close to, 125
  innovation, supply chain for, 60, 63
  integrating speed with, 42
  morale and, 155–157
Telephone interviews, competitive intelligence and, 93
Thompson John W., 119